Protocols in Biochemistry and Clinical Biochemistry

Protocols in Biochemistry and Clinical Biochemistry

BUDDHI PRAKASH JAIN
Assistant Professor, Department of Zoology, School of Life Sciences, Mahatma Gandhi Central University Bihar, Motihari, India

SHYAMAL K. GOSWAMI
Professor, School of Life Sciences, Jawaharlal Nehru University, New Delhi, India

SHWETA PANDEY
Govt VYT PG Autonomous College, Durg, Chhattisgarh, India

ELSEVIER

ACADEMIC PRESS
An imprint of Elsevier

Academic Press is an imprint of Elsevier
125 London Wall, London EC2Y 5AS, United Kingdom
525 B Street, Suite 1650, San Diego, CA 92101, United States
50 Hampshire Street, 5th Floor, Cambridge, MA 02139, United States
The Boulevard, Langford Lane, Kidlington, Oxford OX5 1GB, United Kingdom

Notices

Knowledge and best practice in this field are constantly changing. As new research and experience broaden our
understanding, changes in research methods, professional practices, or medical treatment may become
necessary.

Practitioners and researchers must always rely on their own experience and knowledge in evaluating and using
any information, methods, compounds, or experiments described herein. In using such information or
methods they should be mindful of their own safety and the safety of others, including parties for whom they
have a professional responsibility.

To the fullest extent of the law, neither the Publisher nor the authors, contributors, or editors, assume any
liability for any injury and/or damage to persons or property as a matter of products liability, negligence or
otherwise, or from any use or operation of any methods, products, instructions, or ideas contained in the
material herein.

Library of Congress Cataloging-in-Publication Data
A catalog record for this book is available from the Library of Congress

British Library Cataloguing-in-Publication Data
A catalogue record for this book is available from the British Library

ISBN 978-0-12-822007-8

For information on all Academic Press publications
visit our website at https://www.elsevier.com/books-and-journals

Publisher: Andre Gerhard Wolff
Acquisitions Editor: Peter B. Linsley
Editorial Project Manager: Allison Hill
Production Project Manager: Kiruthika Govindaraju
Cover Designer: Alan Studholme

Typeset by SPi Global, India

Working together
to grow libraries in
developing countries

www.elsevier.com • www.bookaid.org

Contents

Techniques

1. Centrifugation
2. Electrophoresis
3. Spectrophotometer
4. Chromatography
5. Titration

CENTRIFUGATION

In biology, centrifugation is a technique that utilizes the centrifugal force for the separation of the biological contents from the liquid medium. The separation of the contents depends on their shape, size, density, and the viscosity of the medium. Analytical centrifugation is used for the analysis of the purified macromolecules and was first developed by Svedberg in the late 1920s. Preparative centrifugation deals with the actual separation of biological contents.

Application of preparative centrifugation: Subcellular fractionation, affinity purification of membrane vesicles, etc.

Application of analytical centrifugation: Determination of the purity of macromolecules and their relative molecular mass, detection of changes in conformation, and ligand-binding studies.

The process utilizes the principle that a particle, in a liquid medium, will sediment much faster when placed under the influence of centrifugal force. While moving through a viscous medium, they experience a frictional force. This force acts in the opposite direction to its sedimentation.

Relative centrifugal force:

$$F = M\omega^2 r$$

where
M = mass of particle
r = radius of the rotor
ω = angular velocity
Angular velocity in rad/s:

$$\omega = \frac{2\pi s}{60}$$

where
s = speed of the rotor (revolution/min)
ω = angular velocity (rad/s)
Applied centrifugal force (G) is

$$G = \omega^2 r$$

or

$$G = \frac{4\pi^2 s^2 r}{3600}$$

The unit of G is cm/s.
Relative centrifugal force is

$$RCF = \frac{G}{g}$$

or

$$RCF = \frac{4\pi^2 s^2 r}{(3600)(981)}$$

or

$$RCF = 1.12 \times 10^{-5} s^2 r$$

According to the Stokes' law, when a particle moves through a viscous medium, it experiences a frictional force:

$$F = 6\pi r_p \eta v$$

where
η = viscosity of the medium
v = velocity of the particle
r_p = radius of the particle

Also, when particle sediment, it feels an upward force. This force will be equal to the weight of water displaced.

$$F = \frac{4}{3}\pi r_p^3 \left(\rho_p - \rho_m\right)\omega^2 r$$

where
ρ_p = density of the particle
ρ_m = density of the medium
r = distance of the particle from the medium
r_p = radius of the particle
ω = angular velocity

A nonhydrated spherical particle will accelerate under centrifugal force till the net force of sedimentation becomes equal to the frictional force.

Or

$$\frac{4}{3}\pi r_p^3 \left(\rho_p - \rho_m\right)\omega^2 r = 6\pi r_p \eta v$$

Thereafter, the particle will sediment at a constant velocity,

$$v = \frac{dr}{dt} = \frac{2}{9} r_p^2 \frac{\left(\rho_p - \rho_m\right)\omega^2 r}{\eta}$$

or the time of sedimentation is

$$t = \frac{9}{2} \frac{\eta}{\omega^2 r_p^2 \left(\rho_p - \rho_m\right)} \ln \frac{r_b}{r_t}$$

where

t = sedimentation time

η = viscosity of the medium

r_p = radius of the particle

r_b = radial distance from the center of rotation to the bottom of the tube

r_t = radial distance from the center of rotation to the liquid meniscus

ρ_p = density of the particle

ρ_m = density of the medium

ω = angular velocity

Sedimentation rate of a particle is expressed as its sedimentation coefficient, s,

$$s = \frac{v}{\omega^2 r}$$

The sedimentation rate per unit centrifugal field can be obtained in various mediums as well as various temperature, therefore the experimental values of the sedimentation coefficient corrected and expressed as sedimentation constant in water as media at 20°C and are denoted as $S_{20,w}$. For biological particles, it is very low and can be expressed as the Svedberg unit, S. One Svedberg unit is equal to 10^{-13} s (Fig. 1).

Types of Centrifuges

There are various types of centrifuges, which differ from each other in the following ways:
1. The maximum speed that can be attained.
2. The maximum volume of samples can be centrifuged.
3. The presence or the absence of a vacuum.
4. The presence or the absence of a refrigeration system, etc.

Some commercially available centrifuges are
1. Large capacity low-speed preparative centrifuges.
2. Refrigerated high-speed preparative centrifuges.
3. Small-scale laboratory microfuges.
4. Large-scale clinical centrifuges.
5. Analytical ultracentrifuges.
6. Preparative ultracentrifuges.

Types of Rotor

1. Fixed angle rotor.
2. Vertical rotor.
3. Swinging bucket rotor.

Types of Centrifugation

1. Differential centrifugation: In this method, the components of various size, shape, and density are separated according to the difference in their sedimentation rate. In this method, the centrifugal field is gradually increased by increasing the speed of the rotor. The medium of separation is homogenous in this case.
2. Density gradient centrifugation: In this method, the particles are separated according to their density in a medium that is nonhomogenous in its density. The density of the medium increases toward the bottom of the centrifuge tube. These can be of two types

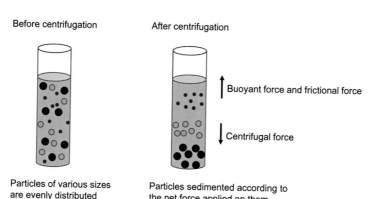

FIG. 1 Pictorial representation of forces involved in centrifugation.

of gradients—continuous and discontinuous. It has two types of variations:

a. Rate zonal centrifugation: The gradient used has a maximum density less than that of the least dense particle. The density gradient is reasonably shallow. Centrifugation is done at a comparatively low speed for a short time. After centrifugation, the particles form discrete zones depending on their sedimenting rate. The centrifugation is terminated before any of the zones reach the bottom of the tube. It is used to separate particles that differ in their size but not density.

b. Isopycnic centrifugation: In this method, the gradient used has a maximum density greater than the densest particle. The particles travel various distances and become stationary when their density becomes less than that of the region below it. It is used to separate the particle, which differs in their density.

ELECTROPHORESIS

The migration of charged particles under the influence of the electric field is termed as electrophoresis. It was first performed in 1861 by Quincke. There are various biological macromolecules, for example, nucleotides, nucleic acids, amino acids, peptides, proteins, etc., that possess ionizable groups, which get ionized at a pH and migrate when placed in an electric field. Certain cells such as bacteria and red blood cells also migrate in the electric field. The migration toward anode or cathode depends on the kind of net charge on the molecule.

When a spherical molecule, possessing a charge q, is under the influence of the electric field, then the force acting on it can be described as follows:

$$F = \frac{\Delta E q}{d}$$

where

ΔE = potential difference between the two electrodes
q = net charge of the molecule
d = distance between the electrodes

While migrating, the particle will also experience a frictional force, opposite to the direction of its migration. This frictional force depends on the size and shape of the molecule, pore size of the medium in which electrophoresis occurs and the viscosity of the medium. Neglecting the pore size of the medium, it can be expressed as

$$F_f = 6\pi r \eta \vartheta$$

where

F_f = frictional force
r = radius of the particle
η = viscosity of the medium
ν = velocity of the particle

Remember, it is an oversimplified situation, where the pore size of the medium is not considered, and the particle is assumed to be spherical.

Therefore, the velocity of the particle in an electric field is

1. Directly proportional to the electric field applied and charge of the molecule.
2. Inversely proportional to the viscosity and its size.

Factors affecting the electrophoretic mobility:

1. Size of the sample: Bigger the sample size greater is the frictional force.
2. Charge of the sample: Higher the net charge of the particle greater is the mobility.
3. The shape of the particle: A globular protein (compact shape) will migrate faster than a fibrous protein.
4. Electric field: The rate of migration under unit potential gradient is known as mobility of ion. The current in the solution is conducted by ions of buffer and can be expressed as

$$V = IR$$

It shows that if the potential gradient increases, it will increase the current and, therefore, the mobility of an ion. It seems the mobility of an ion increases with the increase in potential gradient. However, during electrophoresis, power is generated in the medium of electrophoresis and is dissipated in the form of heat. This heat generation will have the following effects:

o Increased rate of diffusion of buffer ions and sample, which leads to the broadening of the samples.
o Convection current will form, which will mix the separated samples.
o The decrease in viscosity and hence the resistance of the medium.
o Heat-sensitive materials may get affected, e.g., denaturation of proteins can occur.

Therefore, it is a better option to experiment with a constant power supply.

5. The medium used: It should not lead to the adsorption of the sample. Similarly, molecular sieving of the medium affects the migrations of the molecules. The medium generally used are agarose and polyacrylamide.
6. Composition of the buffer: The choice of the buffer depends on the sample to be separated. For example, carbohydrates that are uncharged can be separated using borate buffers, which interact with carbohydrates to form charged complexes.

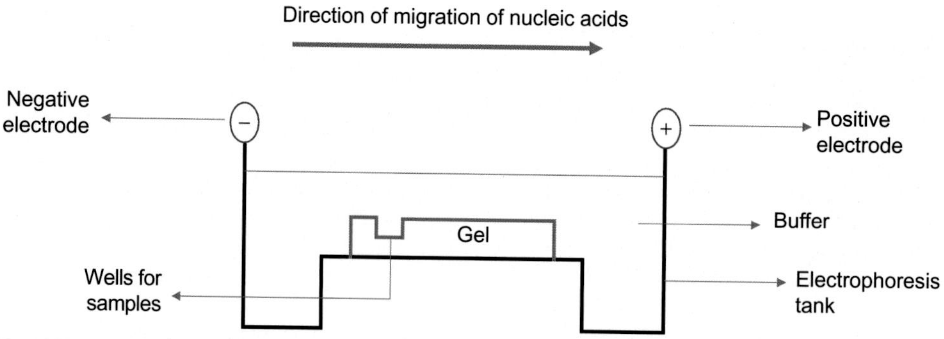

Direction of migration of nucleic acids

FIG. 2 Electrophoresis apparatus.

7. The pH of buffers: pH affects the ionization of the sample molecules and, therefore, their rate and sometimes direction of the sample molecules (in case of ampholytes).
8. Electroendosmosis due to the presence of charged groups on the surface of the support medium too can affect the rate of migration of the particles. Agarose can contain sulfate groups; paper has carboxyl groups and in capillary electrophoresis, the surface of glass comprises silanol (Si-OH) groups (Fig. 2).

SPECTROPHOTOMETER

It is an instrument that measures the amount of light absorbed by a sample. It is mainly used to measure the concentration of a sample.

Light has dual characteristics, particle as well as wave nature. Any object that absorbs light, obeys two laws—Bouguer-Lambert law and Beer's law.

Bouguer-Lambert law—the amount of absorbed by a material is directly proportional to the thickness of the material. It is independent of the intensity of the incident light. It can be represented as

$$\frac{I}{I_0} = e^{-kb}$$

where

I = intensity of the transmitted light

I_0 = intensity of the incident light

k = linear absorption coefficient of the absorbing material

b = thickness of the absorbing material, also known as path length

The above equation can also be written as

$$\ln \frac{I}{I_0} = -kb$$

or

$$\ln \frac{I_0}{I} = kb$$

$$2.303 \log_{10} = kb$$

According to the Beer's law, the amount of light absorbed by a material is directly proportional to the number of molecules of the absorbent, or in other words, the concentration of the absorbing solution. This can be represented as

$$\frac{I}{I_0} = e^{-k'c}$$

or

$$2.303 \log_{10} = k'C$$

Both equations can be combined as

$$\log_{10} \frac{I_0}{I} = abC$$

where

a = combined constant

b = path length

C = concentration of the absorbing material

This equation is Beer-Lambert law, and it states that the amount of the light absorbed is directly proportional to the concentration as well as path length of the absorbing material. I/I_0 is known as transmittance, it is the amount of light that escape absorption by the material. I_0/I is known as absorbance or optical density (OD).

It is easier to use absorbance as an index of concentration as it varies linearly with concentration, while transmittance varies with a concentration in a nonlinear manner.

Absorbance (A) can also be written as

$$A = \log_{10} \frac{I_0}{I} = abC$$

where a is the absorptivity or absorption coefficient or extinction coefficient.

Specific absorption coefficient—when path length is expressed in cm and concentration in g/L, then the absorption coefficient is termed as a specific absorption coefficient.

Molar absorption coefficient—when path length is expressed in cm and concentration in mol/L, then the absorption coefficient is termed as the molar absorption coefficient and is denoted as a_m.

$$a_m = a_s \times Mw \text{ (molecular weight)}$$

For a given compound, if the solvent and wavelength are defined, then molar absorptivity is a physical constant. Molar absorptivity of some compounds are listed as below:

Compound	Solvent	λ_{max}	$a_m \times 10^{-3}$
Adenine	Water	260	13.3
NADH	Water	340	6.22
ATP	Water	260	15.4
FAD	Water	445	11.3

When a sample is placed in a cuvette and light is passed through it, we can calculate the concentration of the sample. When path length is constant, then the optical density is directly proportional to the concentration of the sample, it can be written as follows:

$$\frac{OD_1}{C_1} = \frac{OD_2}{C_2}$$

where

OD_1 and OD_2 = optical density of sample 1 and 2, respectively

C_1 and C_2 = concentration of samples 1 and 2, respectively

Deviations from Beer-Lambert law:

1. When the concentration of the reagent is high: It may lead to dimerization or polymerization of the reagent. OD of monomer differs from polymers. High concentration may also lead to aggregation leading to the formation of aggregates, which scatter light.
2. Temperature: On heating, the solvents may expand. The change in the degree of solubility, dissociation/association of solutes, and hydration of solutes may vary according to the temperature. These changes can lead to variation in absorbance.
3. Turbidity: A turbid solution absorbs more light.
4. Sample instability: Some colored complexes are unstable and their intensity can increase or decrease with time. For example, the ANSA method of phosphate determination.
5. Fluorescence: Some solutes, especially drugs or their intermediates fluoresce, and their fluorescent intensity are also detected by the spectrophotometer.

Absorption spectrum—it is the pattern of energy absorption by a substance, when, the light of varying wavelength passes through it. It is a unique characteristic of the substance as every substance is made up of molecules and each element/ion of the molecule has unique arrangements of an electron in their orbits/orbitals. When light is passed through them, they absorb energy, according to their electronic configuration and their electrons get excited. These electrons get promoted to a higher energy level. According to the quantum theory, the electron gets excited only after accepting the radiation, which has exact quantized energy that can push the electron to a permitted energy level. The wavelength at which it absorbs the maximum amount of light is λ_{max}. When the concentration of any substance is increased, the absorbance at all wavelength increases, although the change in absorbance per unit change in concentration is maximum at λ_{max}. Therefore, during a quantitative experiment, absorbance of a compound is measured at λ_{max}.

Factors that can affect the absorption spectra are polarity (affect the transition states of the electrons) and pH of the solvent (it can change the ionization state of the molecules) as well as the relative orientation of the neighboring absorbing groups.

Some isolated covalently bonded groups characteristically absorb the light in esters, carbonyl, and nitrile group of ethylenic or acetylenic groups. Similarly, there are auxochromes that themselves don't act as chromophores, but their absence or presence shifts the absorption spectrum toward longer wavelengths. They are also known as color enhancers, e.g., OH, OR, SH, NH_2 groups. They can share the nonbonding electrons by extending the conjugation.

There are different conditions like change in the polarity of the solvent or the presence of chromophore, etc., which can shift the absorption spectrum in four different directions and result in the following shifts:

1. Bathochromic shift: Due to the presence of auxochrome, the shift is toward the higher wavelength. This is also known as the redshift.
2. Hypochromic shift: The shift is toward shorter wavelength, due to the removal of conjugation or the change in the polarity of the solvent. This is also known as a blueshift.
3. Hyperchromic shift: This results in an increase in the intensity of the absorption maximum. It results in a higher extinction coefficient. This occurs mostly due to the presence of auxochrome.

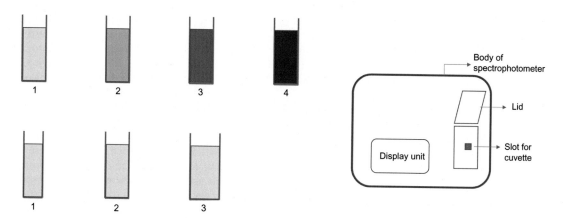

Absorbance of a solution depends on its concentration and path length: (a) As the absorbance depends on the concentration, the absorbance of the tube 4 will be highest and that of the tube 1 will be lowest. (b) (a) As the absorbance depends on the pathlength, the absorbance of the tube 3 will be highest and that of the tube 1 will be lowest.

FIG. 3 Spectrophotometer.

4. Hypochromic shift: The shift of absorption maximum to a lower value. It occurs due to the introduction of the groups that is appropriate result in the distortion of the geometry of the absorbing molecule (Fig. 3).

CHROMATOGRAPHY

The first account of chromatography was given by Michael Tswett in 1906. In the experiment, chlorophyll was separated from a mixture of plant pigments, the colored pigments and chlorophyll get separated forming distinct colored bands. Hence, the term chromatography was given to the separation technique. It is a method of separation of an analyte between two immiscible phases. One of the phases remains stationary, while the other is mobile. The mobile phase usually moves over the surface of the stationary phase. It can also percolate through the interstices of the stationary phase. A sample is introduced in the system with the mobile phase and the components of the sample interact with two phases in a differential manner, depending on their properties. Each component migrates at a different rate based on the rate of interactions between the components of the sample and each of the phases. The component showing the least interaction with the mobile phase but strong interaction with the mobile phase will migrate slowly. The separation is done by the virtue of the differential movement of the components of the sample.

The term partition coefficient or distribution coefficient (K_d) is used to describe the mode of distribution of a compound between two immiscible phases. At a given temperature, when a sample is allowed to equilibrate itself between two different liquids that are immiscible and are present in equal volumes, the ration in which it does the distribution is known as partition coefficient. It is the concentration of a component present in one phase divided by that in another phase. It can be expressed as

$$K_d = \frac{\text{Concentration in phase } A}{\text{Concentration in phase } B}$$

The effective distribution coefficient is the total amount of the analyte in a phase divided its total amount in another phase. It also considers the volume of the phases, for example, if the distribution coefficient of an analyte between A and B phases is 1, but the total volume of phase A is 5 times more than that of B, then the effective distribution coefficient is of the analyte is 5. Its amount in phase A is 5 times more in A than B. In the technique, the stationary phase (immobilized) can be solid, gel, liquid, or solid/liquid mixture, while the mobile phase is liquid or gas (Fig. 4).

Types of Chromatographic Techniques

Based on the nature of the support that is used to hold the stationary phase, the techniques can be classified as follows:

1. Plane chromatography: The following are the types of plane chromatography:

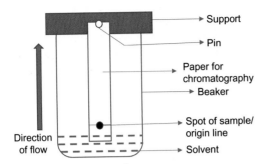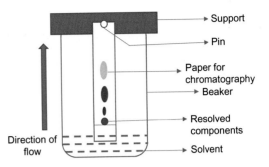

FIG. 4 Chromatography.

a. Paper chromatography: In this technique, the support for the stationary phase is paper, i.e., cellulose. The separation is done by adsorption or partition principle.
b. Thin-layer chromatography: In this technique, the glass plate coated with a layer of silica (partition principle), Kieselguhr, and alumina (adsorption principle).

2. Column chromatography: In this, the stationary phase is packed in a column of glass or metal. The mixture of analytes is applied to the column and the mobile phase passes through it. The mobile phase is termed as eluent. The stationary phase is coated on the matrix and packed in the column or is applied on the walls of the column. The eluent flows through the column and separates the sample according to the partition coefficient of the components in the sample. The components that leave the column individually are termed as an eluate.

Based on the interaction phenomenon between stationary and mobile phase, the technique can be classified in the following types:

1. Partition chromatography: In this, the separation involves a large number of partition steps on the granules of an insoluble hydrated inert substance, e.g., silica gel or starch. The granules are hydrated and hold water in a very tightly bound manner. Hence, water is the stationary phase.
2. Adsorption chromatography: In this chromatography, the components are absorbed with the help of weak electrostatic forces or hydrogen bonding on the surface of the stationary phase.
3. Ion exchange chromatography: In this chromatography, an ion exchanger consisting of inert support covalently coupled with negative (cation exchanger) or the positive (anion exchanger) functional group. The support can exchange the charged functional group with that of the components of the sample.

In the case of an anion exchanger, the negatively charged sample will interact with the stationary phase, while the positively charged and neutral molecules will elute first.

4. Molecular sieve chromatography: In this chromatography, a gel consisting of porous beads is the support medium. The molecules smaller than the pore size will get entrapped in the pore of the gel, while the larger molecules will not and thus get eluted first.
5. Affinity chromatography: In this chromatography, the specificity of an enzyme toward its substrate or substrate analogs. The technique utilizes the specificity of an enzyme for its substrate (also receptor for its agonist, the antibody for an antigen) or substrate analog for the enzyme's (other proteins with biological specificity) separation. A substrate analog is coupled to the gel matrix and the cellular suspension can percolate through. The enzyme which is specific for the substrate analog binds to the gel becoming immobile, while all other components move down and out. The technique has a very high-resolution power.

TITRATION

It is a method to determine the concentration of an identified analyte. It is a volumetric analysis, i.e., a method in which the amount of a solution is determined by measuring its volume. In this process, a volume of a solution of known strength is added to another solution to complete the reaction. There are a few important terms used in the procedure, which can be listed as below:

1. Titer: It is the weight of solution in 1 mL of solution or weight of a substance that will react with 1 mL of solution or is equivalent to 1 mL of solution.
2. Standard solution: A solution of an accurately known concentration is known as the standard solution. They can be of two types—primary standard

solutions—these solutions are prepared directly by mixing the salt and solvent, e.g., N/10 oxalic acid. Secondary standard solutions—these are prepared after standardization through the reaction with primary standard, e.g., N/7 NaOH.

3. End point: The point at which titration is stopped, it is indicated by the change in color of an indicator. The indicator that is used depends on the pH at which the indicator changes the color.

4. Indicator: A high molecular weight substance that indicates the physical and chemical status of a reaction. When dissolved in water, they act as a weak acid or base. The acidic indicator comprises a colored anion, while the basic indicator comprises a colored cation.

Common indicators used are

Indicator	pH Range	Color in Acidic Medium	Color in Basic Medium	Concentration Used
Methyl orange	3.1–4.4	Orange-red	Orange-yellow	0.1%
Methyl red	4.2–6.2	Red	Yellow	0.2% in 95% alcohol
Phenolphthalein	8.2–10.0	Colorless	Pink	1g in 110 mL 95% alcohol and 90 mL of water
Bromocresol green	3.6–5.2	Yellow	Blue	0.04% in alcohol

Types of reaction in titrimetry:

1. Neutralization reaction: The reaction between acid and base.

2. Acidimetry: The reaction in which the amount of base is determined by titrating it against a standard acid solution.

3. Alkalimetry: The reaction in which the amount of acid is determined by titrating it against a standard base solution.

4. Precipitation and complexation reactions: A reaction which forms a precipitate, or a complex approaches the completeness. It occurs due to the removal of ions from the solution.

5. Redox reactions: Reactions involving the transfer of electrons from one molecule to another.

Common equations useful in titration are

$$\text{Volume} \times \text{Normality} = \text{Gram equivalent weight}$$

and

The volume of solution 1 $(V_1) \times$ Normality of solution 1 (N_1) = Volume of solution 2 $(V_2) \times$ Normality of solution 2 (N_2)

In the titration process, the volume of a solution of unknown strength is taken in a burette and this is titrated against a known volume of known strength in a flask. The flask also contains a few drops of indicator. As the reaction reaches the end point, the color of the solution in the flask changes, and the volume of the solution of unknown strength (solution in the burette) is noted. By applying the above formula, the strength of the solution can be obtained.

CHAPTER 1

Solutions and Acids and Bases

The solution is a homogenous mixture of two or more components. Solutions that are made up of two components are binary. The component in which something is dissolved or, in other words, the component which is in larger quantity is known as a solvent; the component which is dissolved or, in other words, which is present in smaller quantity is the solute.

$$Solute + Solvent = Solution$$

Each component may be in any state, i.e., solid, liquid, or gaseous state. The strength of any solution is very important in doing any experiment and it has to be made cautiously. The strength of a solution can be defined as the amount of solute dissolved per unit solution or solvent. It can be expressed in various terms.

MOLARITY (M)

It is the most common way to represent the strength of any solution. It is defined as the no. of moles of a solute per liter of the solution.

$$Molarity = \frac{No. of \ moles \ of \ solute}{Volume \ of \ the \ solution \ in \ liters}$$

For example,

To prepare a solution of 2 M NaOH in 100 mL of solution, we need 2 M of NaOH dissolved in 100 mL of the solution.

1 mol NaOH = 40 g of NaOH
2 mol NaOH = 80 g of NaOH

Now, if in 1000 mL of solution, 80 g of NaOH is dissolved then 2 M solution could be obtained.

For 100 mL of solution, 8 g of NaOH is needed.

But in case of liquid it differs as here density comes into play:

Preparation of 1 M HCl:

MW of HCl is 36.5 g.

So, if 36.5 g of HCl is dissolved in 1 L of solution then 1 M HCl could be obtained.

The specific gravity of HCl is 1.19, which means, 1 mL of HCl weigh 1.19 g.

So, the volume needed is mass/specific gravity or 36.5/1.19.

The volume of HCl needed is 30.67 mL.

But it is true if the conc. HCl is 100%, but in the bottle, the conc. HCl is 37.23%, so the volume needed is 82.38 mL in 1 L of solution.

MOLALITY (M)

A solution containing 1 mol of solute in 1 kg of a solvent is known as 1 molal solution.

$$Molality = \frac{No. of \ moles \ of \ solute}{Weight \ of \ solvent \ in \ kg}$$

For example,

To prepare 1 m solution of NaOH in water, we need 1 mol of NaOH in 1000 g of water.

As specific gravity of water is 1,

1 g water = 1 mL of water.

1 mol of NaOH = 40 g of NaOH.

Now, if we add 40 g of NaOH in 1000 mL of water, it will result in a 1 M solution of NaOH in water.

Note that molarity refers the total volume of solution (solvent + solute) whereas molality refers the weight of the solvent.

NORMALITY

It is the number of gram equivalents of the solute that has been dissolved in 1 L of solution. Gram equivalent or equivalent weight of a given substance is the mass of a substance that combines with or displaces a fixed quantity of other substances. This can be defined as the following:

1. For an element the mass of the element that combines with/displaces 1.008 g of hydrogen, 8.0 g of oxygen, or 35.5 g of chlorine. It can be obtained by dividing the atomic weight of an element by its valency.
2. For acids and bases, it is the mass of acid or base that provides or reacts with 1 mol of H^+.

The unit of equivalent weight is g.

For example, in the case of NaCl: anion is Na^+ and cation is Cl^-. The total charge on anion or cation is 1.

So, the equivalent weight is $(23 + 35.5)/1 = 58.5$.

In the case of $Al_2(SO_4)_3$: anion is two molecules of Al^{3+} and cation is three molecules of $(SO_4)^{2-}$.

Protocols in Biochemistry and Clinical Biochemistry. https://doi.org/10.1016/B978-0-12-822007-8.00009-X

The total charge on either of them is 3×2.
So, the equivalent weight is $342.15/6 = 57$.

$$\text{Normality} = \frac{\text{Gram equivalent of solute}}{\text{Volume of the solution in a liter}}$$

Mass Concentration

Mostly, the macromolecules like nucleic acids, proteins do not have a uniformly defined composition. In that case, their concentration is defined as weight per unit volume or in terms of percentage.

PERCENT

Part of solute dissolved to form a hundred parts of the solution.

For example, 2% NaOH (w/v) means, 2 g of NaOH in 100 mL of solution.

Similarly, 2% HCl means,

2 g of HCl in 100 mL of solution (w/v).
2 g of HCl in 100 g of solution (w/w).
2 mL of HCl in 100 mL of solution (v/v).

MASS FRACTION

Mass fraction of a component in the solution means the mass of the component per unit mass of the solution.

Let component A of the solution has mass W_a and component B has W_b.

Then,

$$\text{Mass fraction of } A = \frac{W_a}{W_a + W_b}$$

Similarly,

$$\text{Mass fraction of } B = \frac{W_b}{W_a + W_b}$$

If a substance is present in a very low amount the term parts per million is used. It is the gram of solute per million grams of the solution.

$$\text{Conc of solute in ppm} = \frac{\text{g or mL of solute}}{\text{g or mL of solution}} \times 10^6$$

Acids and Bases

There are three major theories describing acids and bases.

1. Arrhenius theory

Acid: According to Arrhenius, the compound that ionizes to produce a proton (H^+) when dissolved in water, e.g., HCl.

HCl is strong acid while HNO_2 is a weak acid.

$$HCl \longrightarrow H^+ + Cl^-$$

$$HNO_2 \rightleftharpoons H^+ + NO_2^-$$

Base: The compound that can produce hydroxide ion (OH^-) when dissolved in water, e.g., NaOH.

$$NaOH \rightarrow Na^+ + OH^-$$

2. Bronsted-Lowry theory

According to this theory, any compound that yields proton (H^+) is acid and any compound that accepts the proton is base. So, according to this theory:

Acid		Base
HCl	H^+	Cl^-
H_2SO_4	H^+	SO_4^{2-}
HNO_3	H^+	NO_3^{3-}

Acid and its base are said to be conjugated. A strong acid has a weak conjugate base, and weak acid has a strong conjugate base. Similarly, the weak base has a strong conjugate acid and a strong base has a weak conjugate acid.

3. Lewis theory: According to this theory, acids are the compounds that accept a pair of electrons while bases donate a pair of electrons, e.g., NH_3 is a base while BF_3 is an acid.

Electrolytes

Electrolytes are the compounds that form ions when dissolved in water or any polar liquid, e.g., NaCl. Major biomolecules in our cells, e.g., peptides, amino acids, nucleotides, nucleosides, and nucleic acids, are weak electrolytes and dissociates partially in aqueous solutions. The biological function of most of these biomolecules depends on the pH. The pH of a solution is the measure of the alkalinity or acidity. It is the negative of the logarithm of the molar hydronium (or hydrogen) ion concentration.

$$pH = -\log[H_3O^+] = -\log[H^+]$$

Water is the most important weak electrolyte and partially exists in its ionized form:

$$H_2O \rightleftharpoons H^+ + OH^-$$

The H^+ ion again reacts with water to form hydronium ions (H_3O^+):

$$H^+ + H_2O \rightleftharpoons H_3O^+$$

The equilibrium constant of water at 25°C is 1.8×10^{-16}. So,

$$K_{eq} = \frac{[H^+][OH^-]}{[H_2O]} = 1.8 \times 10^{-16}$$

The molarity of pure water is 55.6 M. So,

$$1.8 \times 10^{-16} = \frac{[H^+][OH^-]}{55.6}$$

or

$$[H^+][OH^-] = 1.8 \times 10^{-16} \times 55.6 = 1 \times 10^{-14} = K_w$$

K_w is also known as the autoprotolysis constant of water. It is specific for 25°C and depends on temperature.

In the case of acids and bases,

$$K_a + K_b = K_w$$

where K_a and K_b are equilibrium constant of acid and base, respectively.

For example, in the case of acetic acid (CH_3COOH)

$$CH_3COOH \rightleftharpoons CH_3COO^- + H^+$$

and

$$K_a = \frac{[CH_3COO^-][H^+]}{[CH_3COOH]} = \frac{[conjugated\ base][H^-]}{[weak\ acid]}$$
$$= 1.75 \times 10^{-5} \text{ and its } pK_a \text{ is } 4.75$$

Further,

$$CH_3COO^- + H_2O \rightleftharpoons CH_3COOH + OH^-$$

and

$$K_b = \frac{[CH_3COOH][OH^-]}{[CH_3COO^-]} = \frac{[weak\ acid][OH^-]}{[conjugated\ base]}$$
$$= 1.77 \times 10^{-10} \text{ and its } pK_b \text{ is } 9.25$$

On multiplying both we get

$$K_a \times K_b = [H^+][OH^-] = K_w = 1 \times 10^{-14}$$

and

$$pK_a + pK_b = pK_w$$

Stronger acid will have a smaller numerical value of pK_a (and larger K_a) than a weaker acid. The smaller the pK_a value, the more it will be in ionized form.

Buffer

A solution that resists change in its pH, upon the addition of acid or base, is known as a buffer solution. Mostly they are weak electrolytes, due to this their ionic status varies with pH. In an aqueous solution, they are either a mixture of a weak acid with its conjugate base or a weak base with its conjugate acid. For the preparation of buffer solution or to estimate its pH Henderson-

Hasselbalch equation is very important. According to the equation, for weak acid:

$$pH = pK_a + \log \frac{[conjugated\ base]}{[weak\ acid]}$$

or

$$pH = pK_a + \log \frac{[ionized\ form]}{[unionized\ form]}$$

Similarly, for a weak base:

$$pH = pK_a + \log \frac{[weak\ base]}{[conjugated\ acid]}$$

or

$$pH = pK_a + \log \frac{[unionized\ form]}{[ionized\ form]}$$

Buffering capacity is the ability of a buffer to resist a change in pH when an acid or base is added to it. It is denoted as β and can be defined as the moles of acid or base which is required to change the pH of the buffer by one unit.

$$\beta = \frac{db}{dpH} = \frac{-da}{dpH}$$

where db and da are the amount (moles) of base and acid, respectively, and dpH is the change in pH.

Problem
Calculate the pH of 0.01 M of a weak acid with $K_a = 1.75 \times 10^{-5}$.

Solution
Let the formula of weak acid be HA.

Then,

$$HA \rightleftharpoons H^+ + A^-$$

$$K_a = \frac{[H^+][Cl^-]}{[HA]} = 1.75 \times 10^{-5}$$

HA will dissociate to form equal amounts of ions H^+ and Cl^-.

Let x be the amount of H^+ forms. Then Cl^- will be x and HA will be $0.01 - x$ M.

$$1.75 \times 10^{-5} = \frac{[x][x]}{[0.01 - x]}$$

$$1.75 \times 10^{-7} - 1.75 \times 10^{-5}x = x^2$$

In this equation we can neglect x as its value is very small, so the above equation can be written as

$$1.75 \times 10^{-7} = x^2$$

or

$$x = 4.18 \times 10^{-4}$$

pH is $-\log [H^+]$, so,

$$pH = -\log(4.18 \times 10^{-4})$$

$$pH = -(-4 + \log 4.18)$$

or

$$pH = 3.38$$

Question 2

The pH of a solution, which is a mixture of acetic acid and sodium acetate, is 5.06. The concentration of free acetic acid in it is 0.1 N and that of sodium acetate is 0.2 M. Calculate pK_a.

Answer 2

According to Henderson-Hasselbalch equation

$$pH = pK_a + \log \frac{[\text{conjugated base}]}{[\text{weak acid}]}$$

In this case,

$$pH = pK_a + \log \frac{[\text{acetate}]}{[\text{acetic acid}]}$$

$$5.06 = pK_a + \log \frac{[0.2]}{[0.1]}$$

$$pK_a = 5.06 - \log [0.2]$$

$$pK_a = 5.06 - 0.301$$

$$pK_a = 4.76$$

In an experiment, the type of the buffer used depends on the pK_a value. In addition to this, the properties of the buffers are also considered in the selection of buffers for experiments. The properties of a few buffers are discussed in the following:

1. Phosphate buffers: They have high buffering capacity, both Na^+ and K^+ are highly soluble in water so any ratio of Na^+ and K^+ can be used to prepare the buffer. Even with low molarity, a solution of high ionic strength can be obtained. However, it is not possible to form a phosphate buffer solution of high buffer capacity with low ionic strength. Another disadvantage of using phosphate buffers is that they may bind polyvalent cations, especially Ca^{2+} ions. They are known to be toxic to mammalian cells. They lack buffering capacity in the range of 7.5–8.0, they get precipitated with ethanol, so cannot be used in DNA and RNA precipitation process.

2. Tris buffer: It is the most used buffer in biochemical experiments. pK_a of tris base is 8, therefore it has a high buffering capacity between 7.5 and 8.5. It does not interfere with most of the biochemical reactions. It has very low toxicity. It is highly polar and thus soluble in aqueous solution; therefore, it cannot diffuse across the biological membrane and cannot affect the intracellular pH. It is not susceptible to salts. It does not absorb any light in the visible or ultraviolet region. However, the dissociation of tris is affected by temperature and concentration of tris, e.g., with a 1° C increase in the temperature, the pH of the solution decreases by 0.03 units and 10 mM and 100-mM tris solution differ in pH by 0.1 unit of pH. The higher the concentration of tris the higher the pH. It reacts with some metal ions like Ca^{2+}, Cu^+, Ni^{2+}, etc. (like phosphate buffers). It also reacts with glass electrode hence can lead to error in pH reading. Tris reacts with various fixatives like glutaraldehyde, formaldehyde, and glyoxal.

3. Carbonated buffer: The disadvantage is that most metal carbonates are insoluble in water. Its pH is sensitive to temperature. It works in the range 10–10.8.

4. Glycylglycine buffer: It is often used in enzymological experiments. Its best working range is 7.5–8.0. It shows no affinity toward divalent cations like Ca^{2+} and Mg^{2+}. It also has very low UV absorbance. However, being a peptide, it can be cleaved by protease and hence cannot be used in reactions involving proteases and crude protein preparation.

5. Triethanolamine buffer: This too is majorly used in enzymological experiments. Like the above buffer, its best working range is 7.5–8.0 and shows no affinity toward divalent cations like Ca^{2+} and Mg^{2+}. It too has very low UV absorbance and unlike glycylglycine buffer, it is not cleaved by proteases. However, the buffer is volatile and is therefore suitable for experiments in which buffer is ultimately removed, e.g., purification experiments.

6. Good buffers: Discovered by Norman Good. These are the best suitable buffers for a variety of molecular biology work. They are not toxic, do not precipitate divalent ions, do not absorb in the UV range, and their pH is not sensitive to the change in temperature. Their name is long so is written in abbreviations, e.g., MOPS [3-(N-morpholino) propane sulfonic acid]—working range 6.5–7.9 and PIPES (1,4-piperazinediethanesulfonic acid)—working range 6.4–7.2.

CHAPTER 2

Carbohydrate

A. Qualitative test for carbohydrates
  1. Determination of the presence of carbohydrates in the given sample by Molisch's test.
  2. Determination of the presence of reducing carbohydrates in the given sample by picric acid test.
  3. Determination of the presence of reducing carbohydrates in the given sample by Fehling's test.
  4. Determination of the presence of reducing carbohydrates in the given sample by Benedict's test.
  5. Determination of the presence of reducing carbohydrates in the given sample by Tommer's test.
  6. Determination of the presence of reducing carbohydrates in the given sample by Nylander's test.
  7. Distinguish between monosaccharides and reducing disaccharides by Barfoed's test.
  8. Distinguish between aldose and ketose by Seliwanoff's test.
  9. Determination of pentose sugar by Bial's test.
  10. Determination of the presence of galactose in a test sample by the Mucic acid test.
  11. Differentiation between the presence of ketohexose and aldohexose in the test sample by Foulger's test.
  12. Determination of the presence of reducing carbohydrates in the given sample by Fearon's methylamine test.
  13. Confirmation of the presence of reducing carbohydrates in the given sample by osazone formation test.
  14. Determination of the presence of polysaccharides in the given sample by iodine test.
B. Quantitative test for carbohydrates
  15. Quantitative estimation of carbohydrates in the given sample by anthrone test.
  16. Quantitative estimation of reducing carbohydrates in the given sample by 3,5-dinitro salicylic acid (DNSA) test.
C. Other tests for carbohydrates
  17. Extraction and estimation of glycogen from the liver and muscle of the well-fed and starved rat.
  18. Separation and identification of sugars present in fruit juices using thin-layer chromatography (TLC).
  19. Extraction and analysis of soluble carbohydrates from plants.

QUALITATIVE TEST FOR CARBOHYDRATES

Definition

Determination of the presence of carbohydrates in the given sample by Molisch's test.

Rationale

All types of carbohydrates—monosaccharides (except triose and tetrose), disaccharides, and polysaccharides—get dehydrated upon treatment with conc. sulfuric acid or hydrochloric acid and produce an aldehyde. This aldehyde condenses with two molecules of α-naphthol to form a purple-colored ring.

$$\text{Pentose sugar} \xrightarrow{\text{conc.} H_2SO_4 \text{ or HCl}} \text{Furfural}$$

$$\text{Hexose sugar} \xrightarrow{\text{conc.} H_2SO_4 \text{ or HCl}} 5-\text{Hydroxymethylfurfural}$$

Both aldehydes (furfural and 5-hydroxymethylfurfural) condense with Molisch's reagent (10% α-naphthol in ethanol). Other phenols such as thymol and resorcinol also give a colored product. This test results in the purple-colored product by all carbohydrates that are larger than tetrose. Nucleic acid and glycoproteins too give a positive test. In this test, oligosaccharides or polysaccharides hydrolyzed to form monosaccharides, which ultimately react as above.

$$\text{Furfural}/5-\text{hydroxymethylfurfural} \xrightarrow{\text{Molisch's reagent}} \text{Purple} - \text{colored dye}$$

Materials, equipment, and reagents

A. **Reagents**: Molisch's reagent (10% α-naphthol in ethanol), test sample (sugar solution), concentrated H_2SO_4.
B. **Glassware**: Test tube, test tube holder, dropper.

Protocol

1. Take 2 mL of the sample solution in a test tube.
2. To this add 2–3 drops of Molisch's solution.
3. Add 1 mL of conc. H_2SO_4 or HCl along the sides of the test tube, so that two layers are formed.

Protocols in Biochemistry and Clinical Biochemistry. https://doi.org/10.1016/B978-0-12-822007-8.00007-6

Analysis and statistics

The purple ring indicates the presence of sugar/carbohydrates.

Precursor techniques

1. Molisch's reagent: Add 10g of α-naphthol to 95% ethanol.
2. Sugar solution: If sugar is in powdered form add 1g of sugar to 100 mL of water.

Safety considerations and standards

1. Concentrated acids should be handled very carefully, one must wear gloves while handling it.
2. Always add acid to water.

Pros and cons

Pros	Cons
The easy and primary method to test the presence of carbohydrates	Trioses and tetrose do not give positive result of this test

Alternative methods and protocols

DNS method, anthrone method.

Summary

1. The carbohydrates get dehydrated to form an aldehyde when treated with concentrated acid.
2. The aldehyde formed by dehydration is condensed with Molisch's reagent to form a purple-colored ring.
3. Nucleic acids and other biomolecules that contain sugars too yield color in the test.

MOLISCH TEST

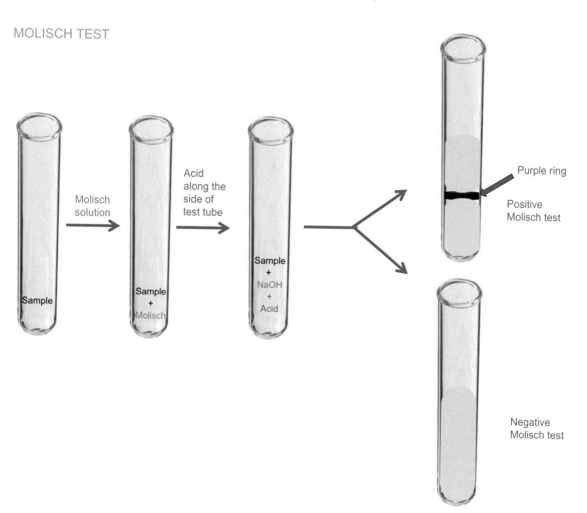

Definition

Determination of the presence of reducing carbohydrates in the given sample by picric acid test.

Rationale

Reducing sugars react with picric acid, i.e., 2,4,6-trinitrophenol (yellow-colored solution) in alkaline solution and reduce it to picramic acid, i.e., 2-amino-4,6-dinitrophenol (red-colored solution.

$$\text{Picric acid (yellow)} \xrightarrow{\text{Reducing sugars alkaline solutions}} \text{Picramic acid (red)}$$

The sugars which have a free aldehyde or ketone group, in other words if the aldose C1 and ketose C2 are not involved in any bond formation, are known as reducing sugars. The free aldehyde or ketone group enables them to work as reducing agents. All monosaccharides are reducing sugars while disaccharides can be classified as reducing or nonreducing.

Picric acid is reduced under the alkaline condition in the presence of reducing sugar. This results in the change of color of the solution from yellow to red.

Materials, equipment, and reagents

A. **Reagents**: Test sample (sugar solution), sodium carbonate (10%) solution, saturated picric acid solution.
B. **Glassware**: Test tube, test tube holder, dropper.
C. **Instrument**: Burner.

Protocols

1. Take 1 mL of test solution in a test tube.
2. To this add 1 mL of saturated picric acid followed by the addition of 0.5 mL of sodium carbonate solution.
3. Mix the above gently and heat for 10–20 s.

Analysis and statistics

Reducing sugar is present in the solution.

Precursor techniques

1. Sugar solution: If sugar is in powdered form add 1 g of sugar to 100 mL of water.
2. Saturated picric acid solution: 1.3 g of picric acid dissolved in 100 mL of water.
3. 10% sodium carbonate solution: Add 10 g of Na_2CO_3 to 100 mL of distilled water.

Safety considerations and standards

1. Picric acid is a highly acidic solution with pH 2 and hence should be handled very carefully.

Pros and cons

Pros	Cons
Easy and sensitive experiment	Indicates the presence of reducing sugars only

Alternative methods/procedures

Fehling's test, Benedict's test, Tommer's test.

Summary

1. In the test, reducing sugars reduces picric acid (yellow) to picramic acid (red) in the presence of alkaline conditions.
2. Both monosaccharides and reducing sugars yield a positive result in the test.

PICRIC ACID TEST

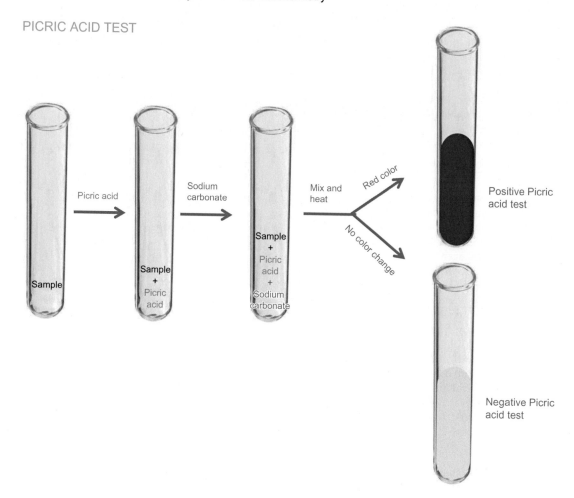

Definition

Determination of the presence of reducing carbohydrates in the given sample by Fehling's test.

Rationale

Reducing sugars get oxidized to acids and the cupric ion gets reduced to cuprous ions, resulting in red- colored precipitate.

The copper ions of Fehling's I solution result in its blue color as they exist in $[Cu(H_2O)_4]^{2+}$ or $[Cu(H_2O)_6]^{2+}$ in the aqueous solution. Upon the addition of Fehling's solution II, copper reacts with tartrate ions and the solution becomes dark blue.

$$[Cu(H_2O)_4]^{2+} + 2C_4H_4O_6^{2-} + 2OH^-$$
$$\text{Copper(aq)} \qquad \text{(Tartrate ion)}$$
$$\rightarrow \quad [Cu(C_4H_3O_6)_2]^4 \quad + 6H_2O$$
$$\text{Copper complexed with tartrate ions}$$

The aldehyde/ketone group is oxidized to acid and Cu^{2+}, which is bound to copper tartrate complex, reduces to copper(I) hydroxide.

Reduction

$$\left[\overset{+2}{Cu}(C_4H_3O_6)_2\right]^4 + 2OH^- + 2e^- \rightarrow 2\overset{+1}{Cu}OH + 4C_4H_3O_6^{2-}$$

This is followed by the dehydration of copper(I) hydroxide and red precipitate of copper(I) oxide is formed.

$$2\overset{+1}{Cu}OH \rightarrow Cu_2^{+1}O \downarrow + H_2O$$

The aldehyde/ketone group is oxidized to acid:

$$RCHO + 2OH^- \rightarrow RCOOH + H_2O + 2e^-$$

This is followed by deprotonation of carboxylic acid:

$$RCOOH + OH^- \rightarrow RCOO^- + H_2O$$

Therefore, the net reaction is

$$2\left[Cu(C_4H_3O_6)_2\right]^{4} + 5OH^- \rightarrow Cu_2O\downarrow$$
$$+ 3H_2O + \left[Cu(C_4H_3O_6)_2\right]^{4-} + RCOO^-$$

In aqueous solution, only less than 1% sugars exist in open-chained form. However, the reaction products, i.e., copper(I) oxide and gluconic acid, are continuously removed as precipitate and gluconic ion in the reaction, therefore the ring form of sugar is converted to the open-chain form till the completion of the reaction.

Materials, equipment, and reagents
A. **Reagents**: Test sample (sugar solution), Fehling's solution A, Fehling's solution B.
B. **Glassware**: Test tube, test tube holder, dropper.
C. **Instrument**: Water bath.

Protocols
1. Take 1 mL of test solution in a test tube.
2. To this add 1 mL of Fehling's solution (A+B, 0.5 mL each).
3. Mix the above gently and keep in a water bath at 80° C for 3–5 min.

Analysis and statistics
Mixing of Fehling's solution, A (blue), and B (colorless) results in a darker blue-colored solution. When this mixture is added to a sugar solution and heated in a water bath and the red precipitate is formed, then it indicates the presence of reducing sugar in the test solution.

Precursor techniques
1. Sugar solution: If sugar is in powdered form add 1 g of sugar to 100 mL of water.
2. Fehling's solution A: Dissolve 3.5 g of cupric sulfate pentahydrate ($CuSO_4 \cdot 5H_2O$) in water to make a final solution of 50 mL.
3. Fehling's solution B: Dissolve 10 g of sodium hydroxide (NaOH) and 17.5 g of potassium-sodium tartrate (Rochelle salt; $KNaC_4H_4O_6 \cdot 4H_2O$) in 50 mL of water.
4. Before the test, both Fehling's solutions A and B should be added in equal amount, as per the required amount.

Safety considerations and standards
1. Fehling's solution B can cause serious eye damage and skin irritation, so it must be handled carefully and one must wear gloves while handling it.

Pros and cons

Pros	Cons
An easy and quick method	Determines the presence of reducing sugars only

Alternative methods/procedures
Benedict's test and Tommer's test.

Summary
1. Fehling's solution A is blue and becomes dark blue when mixed with Fehling's solution B.
2. When the mixture of the above solution is mixed with reducing sugars, they get oxidized to acids and the cupric ion gets reduced to cuprous ions, resulting in red-colored precipitate.

FEHLING TEST

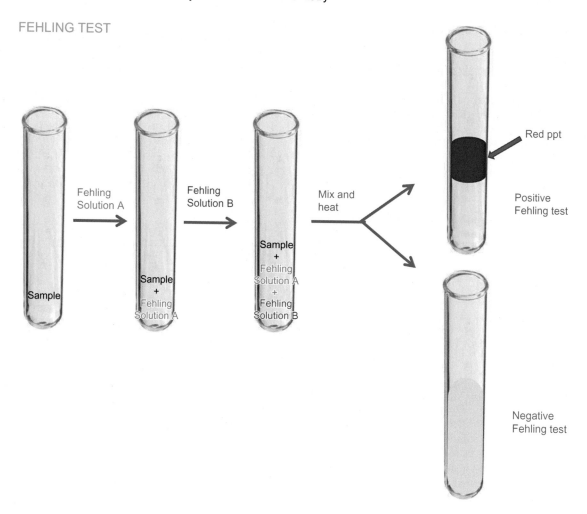

Definition

Determination of the presence of reducing carbohydrates in the given sample by Benedict's test.

Rationale

Benedict's solution contains mild alkali Na_2CO_3. Upon heating in the presence of alkali, reducing sugars are converted into enediol. Enediols are alkenes with a hydroxyl group on each carbon of $C{=}C$ and are very powerful reducing agents. The cupric ions (Cu^{2+}) of Benedict's reagent get reduced to cuprous form (Cu^+) as cuprous oxide (which gets precipitated) by enediols.

$$\text{Reducing sugar} + \text{Cu(citrate)}_2^{2-} \rightarrow \underset{\text{Carboxylate ion}}{\text{RCOO}^-} + \underset{\text{Cuprous oxide}}{\text{Cu}_2\text{O} \downarrow}$$

In Benedict's solution, copper sulfate furnishes cupric ions, sodium carbonate makes the medium alkaline and sodium citrate complexes with cupric ions to prevent them from deteriorating to cuprous ions on storage.

Materials, equipment, and reagents

A. **Reagents**: Test sample (sugar solution), Benedict's reagent.
B. **Glassware**: Test tube, test tube holder, dropper.
C. **Instrument**: Water bath.

Protocols

1. Take 1 mL of test solution in a test tube.
2. To this add 2 mL of Benedict's solution.
3. Mix the above gently and keep in a water bath at 80° C for 5 min.

Analysis and statistics

If the solution turns red, it indicates the presence of reducing sugar in the test solution.

Precursor techniques

1. Sugar solution: If sugar is in powdered form add 1 g of sugar to 100 mL of water.
2. Benedict's solution: Dissolve 17.3 g of sodium citrate and 10 g of sodium carbonate in 80 mL water. In a separate beaker, add 1.73 g of copper sulfate to 10 mL of water. Now, mix both the above solutions and make up the volume up to 100 mL.

Safety considerations and standards

1. Sodium carbonate is a skin irritant, so it must be handled carefully and one must wear gloves while handling it.

Pros and cons

Pros	Cons
Easy and quick method	Determines the presence of reducing sugars only
Benedict modified the Fehling's test, to by producing a single solution, which is more stable and convenient	

Alternative methods/procedures

Fehling's test, Tommer's test.

Summary

1. The reducing sugar gets oxidized and reduces the cupric ions present in Benedict's solution to yield red color under alkaline conditions.
2. The alkaline condition is provided by sodium carbonate.

Definition

Determination of the presence of reducing carbohydrates in the given sample by Tommer's test.

Rationale

Reducing sugars get oxidized to acids and the cupric ion gets reduced to cuprous ions, resulting in the red-colored precipitate.

$$\text{Reducing sugar} + Cu^{2+} \xrightarrow{\text{Alkali}} \underset{\text{Carboxylate ion}}{RCOO^-} + \underset{\text{Cuprous oxide}}{Cu_2O \downarrow}$$

In Tommer's test, copper sulfate furnishes cupric ions; sodium hydroxide makes the medium alkaline. In the presence of alkali, the cupric ions get reduced to cuprous ions forming a red-colored precipitate.

Materials, equipment, and reagents

A. **Reagents**: Test sample (sugar solution), Tommer's reagent.
B. **Glassware**: Test tube, test tube holder, dropper.
C. **Instrument**: Water bath.

Protocols

1. Take 1 mL of test solution in a test tube.
2. To this add 1 mL of Tommer's reagent.
3. Mix the above gently and keep in a water bath at 80° C for 2–3 min.

Analysis and statistics

If a red-colored precipitate is formed, then it indicates the presence of reducing sugar in the test solution.

Precursor techniques

1. Sugar solution: If sugar is in powdered form add 1 g of sugar to 100 mL of water.
2. Tommer's solution: Dissolve 5 g of sodium hydroxide (NaOH) in 90 mL water and 5 g copper sulfate ($CuSO_4$) in 90 mL water separately. Mix both and make the solution to a final volume of 200 mL.

Safety considerations and standards

1. Sodium hydroxide is a skin irritant, so it must be handled carefully and one must wear gloves while handling it.

Pros and cons

Pros	Cons
Easy and quick method	Determines the presence of reducing sugars only

Alternative methods/procedures

Fehling's test and Benedict's test

Summary

1. The reducing sugars reduce the cupric ions to form red precipitate under alkaline conditions provided by sodium hydroxide.

Definition

Determination of the presence of reducing carbohydrates in the given sample by Nylander's test.

Rationale

Reducing sugars get oxidized to acids and reduce bismuth subnitrate to black precipitate of metallic bismuth under alkaline condition.

$$\text{Bismuth subnitrate} + \text{alkali} \longrightarrow \text{Bismuth hydroxide}$$

$$\xrightarrow{\text{Reducing sugar, heat}} \text{Bismuth(s)}$$

In Nylander's test, the bismuth subnitrate is reduced to black-colored metallic bismuth in the presence of alkali and reducing sugar. The reducing sugar gets oxidized to acid.

Materials, equipment, and reagents
A. **Reagents**: Test sample (sugar solution), Nylander's solution.
B. **Glassware**: Test tube, test tube holder, dropper.
C. **Instrument**: Water bath.

Protocols
1. Take 1 mL of test solution in a test tube.
2. To this add 1 mL of Nylander's solution.
3. Mix the above solution gently and keep it in a water bath at 80°C, for 3 min.

Analysis and statistics
If the solution turns black, it indicates the presence of reducing sugar in the test solution.

Precursor techniques
1. Sugar solution: If sugar is in powdered form add 1 g of sugar to 100 mL of water.
2. Nylander's reagent: Dissolve 20 g of potassium hydroxide (KOH) in 150 mL of water. To it dissolve 4 g of bismuth subnitrate [$Bi_5O(OH)_9(NO_3)_4$] and 8 g of potassium sodium tartrate (Rochelle salt, KNa-$C_4H_4O_6\cdot4H2O$) and make the final volume to 200 mL. Heat the mixture at 50°C till all the salts get dissolved. Cool and then filter it.

Pros and cons

Pros	Cons
Easy and quick method	Determines the presence of reducing sugars only

Alternative methods/procedures
Fehling's test, Tommer's test.

Summary
1. Reducing sugars get oxidized to acids and reduce bismuth subnitrate that is present in the reagent to a black precipitate of metallic bismuth under alkaline condition.

Definition
Distinguish between monosaccharides and reducing disaccharides by Barfoed's test.

Rationale
Barfoed's reagent contains cupric acetate, which gets reduced in the presence of monosaccharides and reducing disaccharides to cuprous oxide (red precipitate). Reducing disaccharides take a much longer time than the monosaccharides as the reducing disaccharides hydrolyze first (in acidic medium) and then react with cupric acetate.

$$RCHO + 2Cu^{2+} + 2H_2O \rightarrow RCOOH + 4H^+ + Cu_2O \downarrow$$

Materials, equipment, and reagents
A. **Reagents**: Test sample (sugar solution), Barfoed's reagent.
B. **Glassware**: Test tube, test tube holder, dropper.
C. **Instrument**: Water bath.

Protocols
1. Take 1 mL of test solution in a test tube.
2. To this add 3 mL of Barfoed's solution.
3. Mix the above gently and keep in a water bath at 80°C for 2–5 min.

Analysis and statistics
The formation of red precipitate within 1–5 min indicates the presence of monosaccharides. If it takes 7–12 min for the red precipitate to form then it indicates the presence of reducing disaccharides.

Precursor techniques
1. Sugar solution: If sugar is in powdered form add 1 g of sugar to 100 mL of water.
2. Barfoed's solution: Dissolve 0.33 mol [60 g of $Cu(CH_3COO)_2$ or 65.9 g of $Cu(CH_3COO)_2\cdot H_2O$] of cupric acetate in 1% acetic acid solution.

Safety considerations and standards
1. Acetic acid should be handled carefully, and one must not inhale its fumes.
2. It is very necessary to keep track of time.

Pros and cons

Pros	Cons
Easy and quick method	Nonreducing sugars cannot be detected
Distinguish between monosaccharides and reducing disaccharides	Barfoed's solution is required in comparatively in more quantity than the test sample

Summary
1. The monosaccharides and reducing disaccharides reduce the cupric ions to red precipitate of copper.

2. This test can even distinguish between monosaccharides and reducing disaccharides as they take time to produce a red color of copper.
3. Disaccharides first get hydrolyzed in the acidic medium followed by their reaction with cupric acetate.

Definition

Distinguish between aldose and ketose by Seliwanoff's test.

Rationale

The test is named after the scientist, Theodor Seliwanoff, who devised it. Polysaccharides and oligosaccharides ketose get hydrolyzed in the presence of concentrated acids to yield simple sugar followed by furfural. The dehydrated ketose reacts with resorcinol to produce a deep cherry red-colored product. Aldose may react (at a much slower rate) to form a faint pink-colored product.

$$\text{Ketose} \xrightarrow{[H]^+} \text{Furfural} \xrightarrow{2\ \text{resorcinal}} \text{Deep cherry red color}$$

Materials, equipment, and reagents

A. **Reagents**: Test sample (sugar solution), Seliwanoff's reagent.
B. **Glassware**: Test tube, test tube holder, dropper.
C. **Instrument**: Water bath.

Protocols

1. Take 1 mL of test solution in a test tube.
2. To this add 2 mL of Seliwanoff's solution.
3. Mix the above gently and keep in a water bath at 80–100°C for 1–3 min.

Analysis and statistics

The formation of deep cherry red-colored precipitate within 1–2 min indicates the presence of ketose. If it takes a longer time for a faint pink-colored precipitate to form, then it indicates the presence of aldose.

Precursor techniques

1. Sugar solution: If sugar is in a powdered form add 1 g of sugar to 100 mL of water.
2. Seliwanoff's solution: Dissolve 0.05 mg of resorcinol in 100 mL of 3 N HCl.

Safety considerations and standards

1. Hydrochloric acid is corrosive and must be handled carefully.
2. It is very necessary to keep track of time.

Pros and cons

Pros	Cons
Easy and quick method	
It can detect ketose sugars specifically	

Summary

1. The test is for the differentiation of aldose from ketose sugars.
2. Only ketose sugars get dehydrated in the presence of conc. acid.
3. The dehydrated ketose reacts with resorcinol to produce a deep cherry red color.

Definition

Determination of pentose sugar by Bial's test.

Rationale

The test is named after the scientist, Manfred Bial. Pentose form furfural in acidic medium. This furfural condenses with orcinol in the presence of ferric ions to form a blue-green-colored complex.

$$\text{Pentose sugar} \xrightarrow{[H]^+} \text{Furfural} \xrightarrow{\text{Orcinol, Fe}^{3+}} \text{Blue green product}$$

$$\text{Hexose sugar} \xrightarrow{[H]^+} \text{Hydroxymethyl furfural} \xrightarrow{\text{Orcinol, Fe}^{3+}} \text{Brown} - \text{grey product}$$

In Bial's reagent, hydrochloric acid provides the acidic medium, ferric chloride provides the Fe^{3+} ions, while orcinol condenses with furfural to form the resultant blue-green-colored product. Hexose generally reacts to form brown-gray color product.

Materials, equipment, and reagents

A. **Reagents**: Test sample (sugar solution), Bial's reagent.
B. **Glassware**: Test tube, test tube holder, dropper.
C. **Instrument**: Water bath.

Protocols

1. Take 2 mL of test solution in a test tube.
2. To this add 2 mL of Bial's solution.
3. Mix the above gently and keep it in a water bath at 80–100°C for 3–5 min.

Analysis and statistics

The appearance of blue-green color indicates the presence of pentose and that of brown-grey color indicates the presence of hexose sugar.

Precursor techniques

1. Sugar solution: If sugar is in powdered form add 1 g of sugar to 100 mL of water.

2. Bial's reagent: Dissolve 0.4g orcinol in 200mL hydrochloric acid and 0.5mL of 10% ferric chloride solution.

Safety considerations and standards
1. Hydrochloric acid is corrosive and must be handled carefully.

Pros and cons

Pros	Cons
Pentose and hexose sugars can be distinguished by this easy method	
Various versions of this test are used for the detection of RNA	

Alternative methods/procedures
Tauber's benzidine test.

Summary
1. The test is specific for pentose sugars.
2. Only pentose sugars get converted to furfural in the presence of an acid, which in turn condenses with orcinol and ferric ions to form the colored complex.

Definition
Determination of the presence of galactose in the test sample by the Mucic acid test.

Rationale
The oxidation of monosaccharides by nitric acid yields soluble dicarboxylic acid. On the other hand, oxidation of galactose as well as lactose yields insoluble mucic acid (also called galactaric acid). Lactose is hydrolyzed to glucose and galactose; it is galactose which gives a positive reaction to the test.

Materials, equipment, and reagents
A. **Reagents**: Test sample (sugar solution), conc. nitric acid.
B. **Glassware**: Test tube, test tube holder, dropper.
C. **Instrument**: Water bath.

Protocols
1. Take 2mL of test solution in a test tube.
2. To this add 5mL of conc. nitric acid.
3. Mix the above gently and keep in a water bath at 80–100°C for 1h or till the solution is reduced to 1mL.
4. Scratch the inner wall of the test tube.
5. Let it cool.
6. Observe the crystals formed under the microscope.

Analysis and statistics
The formation of crystals indicates the presence of galactose in the test solution.

Precursor techniques
1. Sugar solution: If sugar is in powdered form add 1g of sugar to 100mL of water.

Safety considerations and standards
1. Hydrochloric acid is corrosive and must be handled carefully.

Pros and cons

Pros	Cons
The presence of galactose can be detected	It is a time taking process

Summary
1. The test is specific for galactose and galactose containing sugars like lactose.
2. Lactose forms an insoluble mucic acid when it is heated with conc. nitric acid.

Definition
Differentiation between the presence of ketohexose and aldohexose in the test sample by Foulger's test.

Rationale
The dehydration of carbohydrates in the presence of conc. H_2SO_4 yields furfural. The furfural reacts with stannous chloride in the presence of urea and gives a colored complex.

$$\text{Aldohexose} \xrightarrow{[H]^+} \text{Hydroxy methyl furfural} \xrightarrow{SnCl_2,\,urea} \text{Green} - \text{blue product}$$

$$\text{Ketohexose} \xrightarrow{[H]^+} \text{Hydroxy methyl furfural} \xrightarrow{SnCl_2,\,urea}$$
$$\text{Yellow or olive green color}$$

Materials, equipment, and reagents
A. **Reagents**: Test sample (sugar solution), Foulger's reagent.
B. **Glassware**: Test tube, test tube holder, dropper.
C. **Instrument**: Bunsen burner.

Protocols
1. Take 0.5mL of test solution in a test tube.
2. To this add 3mL of Foulger's reagent.
3. Mix the above gently and boil for 45s.

Analysis and statistics

The presence of ketohexose in the test solution is indicated by green-blue color, while the presence of aldohexose results in yellow to olive green color. (Although this test is used to distinguish between ketohexose and aldohexose; pentoses too react with the reagent and yield yellow color.)

Precursor techniques

1. Sugar solution: If sugar is in powdered form add 1 g of sugar to 100 mL of water.
2. Foulger's reagent: Add 40 g urea to 80 mL of 40% H_2SO_4. To this add 2 g of stannous chloride and boil it vigorously till it becomes clear. Cool it and then make up the volume up to 100 mL by adding 40% H_2SO_4.

Safety considerations and standards

1. Sulfuric acid is corrosive and must be handled carefully.
2. The process of boiling should be done carefully.

Pros and cons

Pros	Cons
Easily distinguish between aldohexose and ketohexose	Other carbohydrates too give similar color

Summary

1. The test is to distinguish between ketohexose and aldohexose.
2. Both get dehydrated in the presence of conc. sulfuric acid to hydroxymethylfurfural. But react with stannous chloride to give different colored complexes.

Definition

Determination of the presence of reducing carbohydrates in the given sample by Fearon's methylamine test.

Rationale

Reducing carbohydrates undergo hydrolysis in alkaline conditions to form enediol. The enediol forms a red-colored product after reacting with methylamine hydrochloride. Enediols are alkenes with a hydroxyl group on each carbon of C=C and are very powerful reducing agents.

$$\text{Reducing sugar} \xrightarrow{\text{Alkaline solutions}} \text{Enediol} \xrightarrow{CH_3NH_2 \cdot HCl} \underset{\text{- colored product}}{\text{Red}}$$

Materials, equipment, and reagents

A. **Reagents**: Test sample (sugar solution), sodium hydroxide (10%) solution, 0.2% methylamine hydrochloride.

B. **Glassware**: Test tube, test tube holder, dropper.
C. **Instrument**: Bunsen burner.

Protocols

1. Take 1 mL of test solution in a test tube.
2. To this add 2 mL of methylamine hydrochloride, followed by the addition of 0.2 mL of sodium hydroxide solution.
3. Mix the above gently and heat at 56°C for 30 min or at 100°C for 5 min.

Analysis and statistics

The appearance of red color in the solution indicates the presence of reducing sugar in the test sample.

Precursor techniques

1. Sugar solution: If sugar is in powdered form add 1 g of sugar to 100 mL of water.
2. Methylamine hydrochloride solution: Add 0.2 g of methylamine hydrochloride ($CH_3NH_2 \cdot HCl$) to 100 mL of water.
3. 10% sodium hydroxide solution: Add 10 g of NaOH to 100 mL of distilled water.

Safety considerations and standards

Methylamine hydrochloride can cause skin and eye irritation; when inhaled can irritate the lungs, so it should be handled carefully.

Pros and cons

Pros	Cons
Easy and sensitive experiment	Indicates the presence of reducing sugars only

Alternative methods/procedures

Fehling's test, Benedict's test, Tommer's test.

Summary

1. In this test, reducing sugars are treated with methylamine hydrochloride in alkaline conditions.
2. The sugars get hydrolyzed in the presence of alkali and form enediol.
3. This enediol reacts with methylamine hydrochloride to form a colored product.

Definition

Confirmation of the presence of reducing carbohydrates in the given sample by osazone formation test.

Rationale

Osazones are carbohydrate derivatives that are formed when sugars react with an excess of phenylhydrazine. They are colored crystalline compounds and can be detected under the microscope. Each sugar forms a characteristic crystal. Reducing sugars react with one molecule of phenylhydrazine hydrochloride to form phenylhydrazone hydrochloride, this again reacts with another molecule of phenylhydrazine hydrochloride to give a keto derivate. Finally, the keto derivative reacts with the third molecule of phenylhydrazine hydrochloride.

$$\text{Sugar} + \text{Phenylhydrazine hydrochloride}$$
$$\rightarrow \text{Sugar phenylhydrazone} + H_2O$$

$$\text{Sugar phenylhydrazone} + 2(\text{phenylhydrazine hydrochloride})$$
$$\rightarrow \text{Osazone} + C_6H_5NH_2 + NH_3 + H_2O$$

Materials, equipment, and reagents
A. **Reagents**: Test sample (sugar solution), sodium acetate, phenylhydrazine, glacial acetic acid.
B. **Glassware**: Test tube, test tube holder, dropper.
C. **Instrument**: Water bath.

Protocols
1. Take 5 mL of test solution (2%) in a test tube.
2. To 5 mL of the test solution add 1 g of phenylhydrazine mixture and 2 drops of glacial acetic acid.
3. Mix the above gently and heat in a boiling water bath for approximately 30 min or till the formation of osazone crystals, whichever is earlier.
4. Cool it down and take 2 drops of sugar crystals on a glass slide, put a coverslip on it, and examine it under a microscope.

Analysis and statistics
Osazone crystals of various shapes can be observed under a microscope. The characteristics of the crystals are summarized as follows (Shah, 2016):

Name of sugar	Shape of crystal
Glucose	Needle
Fructose	Needle
Mannose	Needle
Galactose	Balls with a thorny edge
Arabinose	Dense ball needle
Xylose	Fine long needle
Maltose	Sunflower
Lactose	Cotton ball

Precursor techniques
1. 2% Sugar solution: If sugar is in powdered form add 2 g of sugar to 100 mL of water.
2. Phenyl hydrazine mixture: It is prepared by adding 1 part of phenylhydrazine hydrochloride to 2 parts of sodium acetate.

Safety considerations and standards
Be careful while handling the glacial acetic acid.

Pros and cons

Pros	Cons
It is a confirmatory test for respective sugars	None

Summary
1. This is a confirmatory test for various reducing sugars.
2. Each sugar reacts with an excess of phenylhydrazine to form a characteristic crystal, which can be observed under the microscope.

Definition
Determination of the presence of polysaccharides in the given sample by an iodine test.

Rationale
Polysaccharides, especially starch, on reaction with an aqueous solution of iodine, form colored complexes. When dissolved in an aqueous solution of potassium iodide, the elemental iodine yields a triiodide anion (I_3^-). The triiodide anion form complexes with polysaccharides as the result of an intermolecular charge-transfer complex. The intensity of the color decreases with the increase in temperature and organic solvents.

$$\text{Starch} \xrightarrow{I_3^-} \text{Blue} - \text{black color}$$

$$\text{Glycogen} \xrightarrow{I_3^-} \text{Reddish} - \text{brown color}$$

Materials, equipment, and reagents
A. **Reagents**: Test sample (sugar solution), iodine solution.
B. **Glassware**: Test tube, test tube holder, dropper.

Protocols
1. Take 1 mL of test solution in a test tube.
2. To this add five drops of iodine solution.
3. Mix the above and observe the color obtained.

Analysis and statistics
The appearance of blue-black color in the solution indicates the presence of starch and an intermediate reddish-brown color indicates the presence of glycogen in the test sample.

Precursor techniques
1. Sugar solution: If sugar is in powdered form add 1 g of sugar to 100 mL of water.

2. Iodine solution: Prepare 0.005 N iodine solution in 3% potassium iodide solution. To make this add 3 g of KI to 100 mL of the final solution in water followed by the addition of 0.063 g of iodine to it.

Safety considerations and standards

As glycogen and partially hydrolyzed starch give intermediate color (reddish-brown), it should not be confused with the yellowish-brown color of iodine-potassium iodide.

Pros and cons

Pros	Cons
Used for distinguishing the presence of polysaccharides	It cannot be performed at a very low pH, as it results in hydrolysis of polysaccharides
	Increased temperature and presence of organic solvents decrease the intensity of the resultant color

Summary

1. This test can distinguish between starch and glycogen.
2. Polysaccharides react with an iodine solution to form a unique color.

QUANTITATIVE TEST FOR CARBOHYDRATES

Definition

Quantitative estimation of carbohydrates in the given sample by anthrone test.

Rationale

All types of carbohydrates, monosaccharides (except triose and tetrose), disaccharides, and polysaccharides, get dehydrated upon treatment with conc. sulfuric acid or hydrochloric acid and produce aldehydes. This aldehyde reacts with anthrone to yield a bluish-green-colored complex.

$$\text{Pentose sugar} \xrightarrow{\text{conc.H}_2\text{SO}_4 \text{ or HCl}} \text{Furfural}$$

$$\text{Hexose sugar} \xrightarrow{\text{conc.H}_2\text{SO}_4 \text{ or HCl}} 5 - \text{Hydroxymethylfurfural}$$

Both the aldehydes (furfural and 5-hydroxymethylfurfural) condense with anthrone reagent (10% α-naphthol in ethanol). This test results in a bluish-green-colored product by all carbohydrates, which are larger than tetrose.

$$\text{Furfural/5} - \text{hydroxymethylfurfural} \xrightarrow{\text{Anthrone reagent}} \text{Bluish} - \text{green colored product}$$

Materials, equipment, and reagents

A. **Reagents**: Anthrone reagent (0.2% anthrone in conc. sulfuric acid), test sample (sugar solution), standard solutions.
B. **Glassware**: Test tube, test tube holder, dropper.
C. **Instruments**: Water bath, colorimeter.

Protocol

Preparation of standard curve:

(A) **Stock standard (1 mg/mL of glucose)**:
   Weigh 100 mg of glucose and add 100 mL of distilled water to it.

(B) **Working standard (0.1 mg/mL of glucose)**:
   Dilute 10 mL of stock solution to a final volume of 100 mL by adding distilled water to it.
   1. Take six test tubes and label them.
   2. To these add 0–1 mL of working standard solution and add water to make the final volume to 1 mL.
   3. Simultaneously, take 1 mL of the sample solution in another test tube.
   4. Add 4 mL of anthrone reagent, mix well, and cover each of the test tubes.
   5. Put the tubes in a boiling water bath for 10 min.
   6. Cool the test tubes to room temperature.
   7. Measure the optical density at 620 nm.
   8. Subtract the value of the absorbance of blank from the absorbance value for each of the test tubes.
   9. Plot the standard calibration curve with concentration plotted on the X-axis and optical density on the Y-axis.

Test Tube No.	Blank	1	2	3	4	5	Test Sample
Working solution (mL)	0	0.2	0.4	0.6	0.8	1.0	1.0
Water added (mL)	1	0.8	0.6	0.4	0.2	0	0
Anthrone reagent	4	4	4	4	4	4	4
Heat in boiling water bath for 10 min. Cool it to room temperature							
OD at 620 nm	A_0	A_1	A_2	A_3	A_4	A_5	A_T
Final OD	A_0	A_1-A_0	A_2-A_0	A_3-A_0	A_4-A_0	A_5-A_0	A_T-A_0

Analysis and statistics

Calculate the concentration of the test sample corresponding to the optical density obtained, using the linear standard graph.

Precursor techniques

1. Anthrone reagent: Add 0.2 g of anthrone reagent to 100 mL of conc. sulfuric acid (H_2SO_4).

2. Test sample solution: Dissolve the test sample in 10 mL. The solution should be diluted accordingly so that the optical density of the test sample does not exceed that of test tube no. 5, i.e., 0.1 mg/mL of sugar.

Calculation

The concentration can be obtained using a standard graph. The dilution (if made) of the test sample should be multiplied to obtain the final concentration of the test sample solution.

Safety considerations and standards

1. Concentrated acids should be handled very carefully, one must wear gloves while handling it.
2. Always add acid to water.

Pros and cons

Pros	Cons
An easy method to test the quantity of carbohydrates	Trioses and tetrose do not give positive result of this test
	Not suitable when proteins having a high amount of tryptophan is present, as it too contributes to the production of red color

Alternative methods and protocols

Molisch's method.

Summary

1. It is a quantitative test for carbohydrates, except triose and tetrose.
2. In this test, the aldehyde produced by the treatment of carbohydrates with conc. acid reacts with anthrone reagent and yields colored product based on the initial concentration of the sugars.
3. All the tubes containing carbohydrates are treated simultaneously so that the test sample can be compared with the standards.
4. The reaction is sensitive to temperature; therefore, all tubes should be cooled to room temperature before reading the absorbance.

Definition

Quantitative estimation of reducing carbohydrates in the given sample by 3,5-dinitro salicylic acid (DNSA) test.

Rationale

All types of reducing sugars undergo oxidation to form acids and in turn reduce many reagents like DNSA. Under alkaline conditions, the free aldehyde or ketone group of reducing sugar is oxidized to acids and

yellow-colored DNSA is reduced to orange-red 3-amino-5-nitrosalisyclic acid (ANSA).

$$\text{Reducing sugar} + 3,5-\text{dinitrosalicyclic acid} \xrightarrow{\text{Alkali}} \underset{\text{(yellow)}}{} RCOO^-$$
$$+ \underset{\text{(orange-red)}}{3-\text{amino}-5-\text{nitrosalicylic acid}}$$

Materials, equipment, and reagents

A. **Reagents**: Dinitro salicylic reagent, sodium hydroxide solution (2 mol/L), test sample (sugar solution), standard solutions.
B. **Glassware**: Test tube, test tube holder, dropper.
C. **Instruments**: Water bath, spectrophotometer.

Protocol

Preparation of standard curve:
A. **Stock standard (1 mg/mL of glucose)**:
   Weigh 100 mg of glucose and add 100 mL of distilled water to it.
B. **Working standard (0.25 mg/mL of glucose)**:
   Dilute 10 mL of stock solution to a final volume of 40 mL by adding distilled water to it.
   1. Take six test tubes and label them.
   2. To these add 0–1 mL of working standard solution and add water to make the final volume to 1 mL.
   3. Simultaneously, take 1 mL of the sample solution in another test tube.
   4. Add 1 mL of DNS reagent, mix well, and cover each of the test tubes.
   5. Put the tubes in a boiling water bath for 5 min.
   6. Cool the test tubes to room temperature.
   7. Measure the optical density at 540 nm.
   8. Subtract the value of the absorbance of blank from the absorbance value for each of the test tubes.
   9. Plot the standard calibration curve with concentration plotted on the X-axis and optical density on the Y-axis.

Test Tube No.	Blank	1	2	3	4	5	Test Sample
Working solution (mL)	0	0.2	0.4	0.6	0.8	1.0	1.0
Water added (mL)	1	0.8	0.6	0.4	0.2	0	0
DNS reagent	1	1	1	1	1	1	1
Heat in boiling water bath for 5 min. Cool it to room temperature							
OD at 620 nm	A_0	A_1	A_2	A_3	A_4	A_5	A_T
Final OD	A_0	A_1-A_0	A_2-A_0	A_3-A_0	A_4-A_0	A_5-A_0	A_T-A_0

Analysis and statistics

Calculate the concentration of the test sample corresponding to the optical density obtained, using the linear standard graph.

Precursor techniques

1. Sodium potassium tartrate: Add 30 g of the salt to 100 mL of the final solution in water.
2. 3,5-Dinitrosalicylic acid: Dissolve 5 g of this reagent in 100 mL of water.
3. Dinitro salicylic acid (DNS) reagent: Add 50 mL of (1) to 20 mL of (2) and make the solution to a to a final volume of 1 L.
4. Test sample solution: Dissolve the test sample in 10 mL. The solution should be diluted accordingly so that the optical density of the test sample does not exceed that of test tube no. 5, i.e., 0.25 mg/mL of sugar.

Calculation

The concentration can be obtained using a standard graph. The dilution (if made) of the test sample should be multiplied to obtain the final concentration of the test sample solution.

Safety considerations and standards

1. The test is sensitive to the temperature, so cool down all the samples to room temperature before reading.
2. Sodium hydroxide solution is corrosive and irritates skin and eyes, so it must be handled carefully.

Pros and cons

Pros	Cons
An easy and quick method for determining the amount of reducing sugars	Not suitable for the determination of concentration in a complex mixture of sugars

Summary

1. The test is specific for the reducing sugars, which react with DNS reagent in alkaline conditions to form a colored product, which depend on the concentration of the sugar solution.
2. All the tubes containing reducing carbohydrates are treated simultaneously so that the test sample can be compared with the standards.

OTHER TESTS FOR CARBOHYDRATES

Definition

Extraction and estimation of glycogen from the liver and muscle of the well-fed and starved rat.

Rationale

Glycogen is a storage polysaccharide in animals. It is primarily synthesized and stored in the liver and muscles. In the liver, the level of glycogen is affected by the state of diet. In the normal rat, glucose is converted into glycogen by the process of glycogenesis. This process involves many enzymes like phosphorylase, alpha-1,6-glucosidase, phosphoglucomutase, and glucose-6-phosphatase. During starvation, this glycogen which is stored in the liver is broken down in a process termed as glycogenolysis. This process lasts for 1–2 days; after this, the formation of blood glucose occurs from noncarbohydrate sources in the process of gluconeogenesis. While in muscles, its level remains almost constant and is not dependent on the state of diet. As muscles lack the enzyme glucose-6-phosphatase, which aid in the conversion of glycogen to glucose, the glycogen in muscles cannot get converted to blood glucose directly.

The glycogen is released from tissues by homogenization with TCA or by boiling with 30% KOH. It is then get precipitated using 95% ethanol. Sodium sulfate increases the yield of glycogen as it acts as a coprecipitate.

Materials, equipment, and reagents

A. **Reagents**: Tissue (liver, muscle), 5% trichloro acetic acid, 95% ethanol.
B. **Glassware**: Test tube, test tube holder, dropper.
C. **Instruments**: Centrifuge (table top), mortar-pestle.

Protocols

1. Weigh 5 g of tissue, cut into small pieces, now add an appropriate amount of 5% TCA to it, and grind it in mortar and pestle.
2. Pour the homogenate in a centrifuge tube.
3. Centrifuge it at 3000 rpm for 5 min at 4°C.
4. Transfer the supernatant to a graduated tube, again add 5% TCA and repeat the process of homogenization, and centrifuge as above.
5. Collect the supernatant to the same graduated tube.
6. This can be repeated two more times.
7. Take 1 mL of above and to this add 5 mL of 95% ethanol with blowing to effect proper mixing.
8. Now cap the tubes and allow it to stand overnight at room temperature. Alternatively, it can be kept in a water bath at 37°C for 3 h.
9. After the precipitation is completed, centrifuge the tubes at 3000 rpm for 15 min.
10. Discard the supernatant and dissolve the pellet in 5 mL water and reprecipitate by adding 10 mL of 95% ethanol.

11. Recentrifuge the tubes at 3000 rpm for 15 min. Discard supernatant.
12. Dissolve pellets in 2 mL of water.

Alternative method: Using KOH
Materials, equipment, and reagents
A. **Reagents**: Tissue (liver, muscle), 30% KOH solution, saturated Na_2SO_4, 95% ethanol, 1.2 M HCl, 0.5 M NaOH.
B. **Glassware**: Test tube, test tube holder, dropper.
C. **Instrument**: Centrifuge (table top).

Isolation of glycogen:
1. Accurately weigh 1.5 g of the liver and skeletal muscle.
2. In a centrifuge tube take 2 mL of KOH, place the tissue in it.
3. Heat in a boiling water bath for 20 min, with intermittent shaking.
4. Let the tube cool in ice.
5. Add 0.2 mL of saturated Na_2SO_4 and mix it.
6. Add 5 mL of 95% ethanol and allow the tube stand on ice for 5 min. The glycogen gets precipitated.
7. Centrifuge the tube at 3000 rpm for 5 min.
8. Discard the supernatant and add 5 mL of water, mix thoroughly (warm it in case it does not dissolve).
9. Add double distilled water to it to make a solution up to 10 mL. This is glycogen solution in water.
10. If the rat is NOT starved and is well fed, then dilute the solution to 100 mL with water.

Hydrolysis of glycogen:
11. Take 1 mL of glycogen solution in a test tube and add 1 mL of 1.2 M HCl to it. Heat in boiling water bath for 2 h.
12. To this add 1 drop of phenolphthalein and then neutralize with 0.5 M NaOH, till the pink color of the indicator turns yellow.
13. Then dilute it to 5 mL.
14. This results in the hydrolysis of glycogen to glucose, which can be determined by the glucose oxidase method (specific for glucose).

Analysis and statistics
The glycogen extracted from tissues is hydrolyzed and glucose concentration, which is an indication of glycogen content in this context, should be higher in well-fed mice.

Precursor techniques
1. 5% TCA: Weigh 5 g of TCA and add to distilled water to a final volume of 100 mL.
2. 95% ethanol: To 95 mL of ethanol add 5 mL of distilled water.
3. 1.2 M HCl: Add 10.6 mL of HCl to a final volume of 100 mL of solution in distilled water.

4. 0.5 M NaOH: 2 g NaOH in a final volume of 100 mL of solution in distilled water.
5. 5. 30% KOH: 30 g KOH in a final volume of 100 mL of solution in distilled water.

Safety considerations and standards
1. Handle all the reagents very carefully.

Pros and cons

Pros	Cons
An easy method to isolate glycogen from tissues	Time taking

Summary
1. This is a method to isolate glycogen from mice by homogenization of the tissue with TCA or boiling with KOH.
2. The glycogen is obtained from the liver; muscles do not yield it as they lack the enzyme glucose-6-phosphatase.

Definition
Separation and identification of sugars present in fruit juices using thin-layer chromatography (TLC).

Rationale
Fruit juices contain a wide variety of sugars. These sugars can be identified accurately by thin layered chromatography. In the experiment, various sugars have varying solubility in a solvent and this property is utilized in TLC. Depending on the solubility, the sugars travel the distance (upward) with the solvent. As the experiment proceeds, the most soluble sugar will travel the longest distance with solvent and reaches the highest point, while the least soluble sugar will travel the shortest distance. The sugars can be detected after the spraying the solution of aniline-diphenylamine followed by heating.

Materials, equipment, and reagents
A. **Reagents**: Thin layered plates of silica gel G, solvent, fruit juices, absolute ethanol, standard sugar solution, aniline-diphenylamine location reagent.
B. **Glassware**: Glass plates, separation chambers.
C. **Instruments**: Oven, spray gun.

Protocols
1. Take 1 mL fruit juice; add 3 mL of ethanol to remove denatured protein.
2. Spot the supernatant in a thin layered plate with standard sugar solutions.
3. Irrigate the plate with the solvent system in ascending direction in a chamber.
4. Let it develop until the solvent front reaches the top of the plate.

5. Draw a line across it and remove the plate.
6. Let the plate dry in the stream of air.
7. Spray the aniline-diphenylamine solution in a fume chamber and heat the plates at 100°C.
8. Note the color developed for each sugar and calculate the R_f value.
9. Compare the color and R_f values of each sugar sample with that of the standards. The same sugar will have the same color and R_f value.

Analysis and statistics
A comparison of color and R_f value indicates the sugars present in the solution, i.e., fruit juices.

Precursor techniques
1. Thin layered plate of silica gel G: Prepare 0.02 M sodium acetate solution by adding 0.164 g of CH_3COONa to a final volume of 100 mL in water. To this solution add silica gel to make a slurry. Pour it in plates approximately 0.25 mm thick, let it dry. It should be activated before the experiment.
2. Solvent: Mix ethyl acetate, isopropanol, water, and pyridine in the ratio 26:14:7:2.
3. Standard sugar solution (1% in sugar in 10% isopropanol): Add 10 mL of isopropanol to make a final volume of 100 mL, to this add 1 g of desired sugar.
4. Location reagent: It could be prepared freshly by mixing 1% aniline, 1% diphenylamine in acetone, and 85% phosphoric acid in the ratio of 5:5:1.

Safety considerations and standards
1. Phosphoric acid should be handled carefully.
2. Do not inhale the solvent.
3. The R_f value should be calculated precisely.

Pros and cons

Pros	Cons
The identification of individual sugar in a mix is possible	

Alternative methods
Other chromatographic techniques.

Summary
1. The fruit juice along with the standard sugar solution is spotted on the silica plate and allowed to travel with the solvent in an upward direction.
2. As the solvent front reaches the highest point, the plate is dried, and spots are detected.
3. The distance traveled by a sugar is the same in a particular solvent and the sugars can be detected by comparing the spots developed by spraying the detection reagent.

Definition
Extraction and analysis of soluble carbohydrates from plants.

Rationale
Soluble carbohydrates are the carbohydrates that dissolve in the aqueous environment of the cells. Under stress and various stimulations, the respective quantities of the soluble sugars and starch get altered. The pool of soluble carbohydrates varies under various environmental conditions. The soluble carbohydrates get extract with ethanol or water. After the removal of soluble sugars, the starch is hydrolyzed by enzymes and then extracted with water.

Materials, equipment, and reagents
A. **Reagents**: 2% amyloglycosidase and 0.5% α amylase in 0.1 M sodium acetate buffer (pH 4.5), and 95% ethanol.
B. **Apparatus**: Volumetric flask, centrifuge tubes, and mortar pestle.
C. **Instrument**: Water bath.

Protocols
1. Excise the desired part of the plant and cut it into small pieces with the help of a sharp, clean scalpel.
2. This can be frozen or heat inactivated to inactivate the enzymes and slow down the metabolic rate thereby preventing the conversion of sugars, or it can proceed for the extraction of sugars.
3. Rapid freezing in liquid nitrogen.
4. Heat inactivation is done at 90°C for 90 min (minimum for 60 min) followed by grinding in mortar and pestle.
5. For the extraction of sugar, weigh 100–400 mg of dry ground or frozen wet sample in a round bottom plastic 50 mL centrifuge tube.
6. Add 15–20 mL of 95% ethanol and cap it with a rubber stopper having a hole, it should be equipped with glass reflux.
7. Mix and place the mixture in a water bath at 85°C. Note the boiling of ethanol and keep it in boiling ethanol for 20 min.
8. Uncap the tubes and centrifuge it at 10,000 g for 10 min.
9. Decant the supernatant in a volumetric flask of suitable capacity and repeat the extraction for three more times. Make up the volume with 95% ethanol. This is an ethanolic extract.
10. Note that ethanol extract can be held for 2 weeks at 4°C. Similar to ethanol, as used in the above extraction, water too can be used at 60°C followed by centrifugation at 20,000–30,000 g for 10 min. The water

extract can be kept for 5 days at 4°C. Since the starch is partially soluble in water, the water extract is not suitable for subsequent starch determination.

11. For starch extraction, take the pellet of ethanolic extract, let the ethanol evaporate completely.
12. Add 10 mL of water to sample (ethanolic extract in a 50-mL centrifuge tube) and mix.
13. To gelatinize the starch, incubate the samples at 90° C with intermittent mixing for 30 min.
14. Let the samples cool at RT.
15. Add 0.2 M of sodium acetate, pH 4.5, 1 mL amylo-glycosidase, and α amylase to it.
16. Incubate at RT for 1–2 h for α amylase to gelatinize the starch, followed by incubation at 55°C for 16–24 h.
17. Let the tubes cool at RT then centrifuge it at 30,000 g for 20 min.
18. Pour the supernatant in a volumetric flask; add 10 mL water to the pellet, mix it well, and incubate it for 10 min at 60°C.
19. Centrifuge it at 30,000 g for 20 min.
20. Keep the supernatant in the same volumetric flask.
21. Repeat the extraction with water twice.
22. The starch extract can be stored at 4°C for up to 5 days.
23. The total sugar content can be determined by the anthrone method.

Analysis and statistics
The soluble sugar extracts can be obtained as both ethanolic extract and water extract. The starch extract is obtained after removing soluble sugars from the ethanolic extract.

Precursor techniques
1. 95% ethanol: Add 95 mL of ethanol to make a final volume of 100 mL in water.
2. 0.2 M sodium acetate solution (pH 4.5): Mix 2.72 g of sodium acetate with 80 mL water, maintain the pH of 4.5, and adjust the volume to 100 mL with water.

Safety considerations and standards
1. The blank and sample should be processed carefully in the same manner.

Pros and cons

Pros	Cons
An easy method for determination of sugars and starch	Time taking process

Alternative methods
Method of Ichimura and Hisamatsu (1999).

Summary
1. The plant material is ground using liquid nitrogen and the soluble carbohydrates are extracted using water or ethanol.
2. The starch is extracted using sodium acetate and enzymatic solution.

Summary of the sugar test.

Sl No.	Test	Substrate	Result
1	Molisch's test	All carbohydrates	The purple ring is formed
2	Picric acid test	Reducing sugars	The yellow color turns red
3	Fehling's test	Reducing sugars	The blue solution to red ppt
4	Benedict's test	Reducing sugars	The blue solution to red ppt
5	Tommer's test	Reducing sugars	The blue solution to red ppt
6	Nylander's test	Reducing sugars	Black ppt
7	Fearon's test	Reducing sugars	The blue solution to red ppt
8	Barfoed test	Mono and disaccharides	Monosaccharides turn blue to red in a short time. Disaccharides turn blue to red take a long time
9	Seliwanoff's test	Ketose and aldose	Ketose give deep cherry red; aldose turns slowly to a faint pink
10	Bial's test	Pentose and hexose	Pentose give blue-green color; hexose give brown-gray color
11	Foulger's test	Ketohexose and aldohexose	Ketohexose gives green-blue color; aldohexose gives yellow or olive green color
12	Mucic acid test	Galactose	Crystals of mucic acid are formed
13	Osazone test	Various sugars	Different types of crystals formed
14	Iodine	Polysaccharides	Starch gives a blue-black color; glycogen gives a red-brown color

CHAPTER 3

Lipid

1. Solubility test for lipids.
2. Emulsion test for lipids.
3. Determination of the degree of unsaturation of fatty acids.
4. Determination of the saponification value of the given fat/oil sample.
5. Determination of the fatty acid value of fat.
6. Acrolein test for the presence of glycerol.
7. Libermann-Buchard test for the detection of cholesterol.
8. Extraction of lipid from leaves (Bligh and Dyer's method and single-extraction method).
9. Extraction of lipid from egg yolk and estimation of phosphorous content in it.
10. Extraction of lipid from tissues (Folch method).

SOLUBILITY TEST FOR LIPIDS

Rationale

Fats are esters of fatty acids. They are amphoteric in nature and harbor a hydrophilic head of glycerol and hydrophobic tail of fatty acids. Other lipids too are hydrophobic. They get dissolved in organic solvents. Triglycerides with a small-chain fatty acids are slightly soluble in polar solvents as water, while those having long-chain fatty acids are insoluble in it and form emulsions. All triglycerides are soluble in diethyl ether, benzene, and chloroform like nonpolar solvents. Their solubility in polar solvents like methanol, ethanol, acetone, etc. increases on heating. Understanding their solubility characteristics aids in the extraction process.

Materials, Equipment, and Reagents

A. **Reagents**: Fatty acids (butyric, palmitic acid), fats, and oils (butter, vegetable oil), solvents. Solvent (water, acetone, ethanol, diethyl ether, chloroform, etc.).
B. **Glassware**: Test tubes, dropper.
C. **Instrument**: Burner.

Protocol

1. Label various test tubes and put 1 g of different lipids in it.
2. To these add 5 mL water.

3. Shake well and check their solubility.
4. Heat it to 50°C for 5 min and again check their solubility.
5. Repeat the test using other solvents and record the solubility.

Analysis and Statistics

The samples that are more soluble in polar solvent have small fatty acids, while the samples having larger fatty acids are less soluble in polar solvents even after heating.

Safety Considerations and Standards

1. Alcohol is highly inflammable so utmost precaution should be applied while heating.

Pros and Cons

Pros	Cons
A preliminary idea about the lipid can be obtained	This test is not specific

Alternative Methods/Procedures

Emulsion test, grease spot test, Sudan test.

EMULSION TEST FOR LIPIDS

Rationale

The emulsion is a mixture of two immiscible liquids. When the sample containing lipids is suspended in ethanol, they get partially solubilize. When the above solution is mixed with water, the formation of cloudy emulsion occurs due to the insolubility of lipids in water. Emulsification agents like soap, bile salts, proteins, etc. aid in the formation of a permanent emulsion.

Materials, Equipment, and Reagents

A. **Reagents**: Food sample, ethanol, distilled water.
B. **Glassware**: Test tubes, dropper.

Protocol

1. Crush the food sample (only in case of solid food sample) and place it in a dry test tube. In case of liquid sample, add a few drops of the sample in a dry test tube.
2. Add 2 mL of ethanol to the samples.

Protocols in Biochemistry and Clinical Biochemistry. https://doi.org/10.1016/B978-0-12-822007-8.00006-4

3. Shake it thoroughly.
4. In case of solid sample, allow the sample to settle down (~5 min) to allow lipid to be extracted and transfer the ethanol to another test tube.
5. Add 2 mL of distilled water to it.
6. Observe the formation of a white-colored emulsion.

Analysis and Statistics

The samples containing lipids will form a white-colored emulsion.

Safety Considerations and Standards

1. In case of solid samples, the crushing of the samples should be done properly.

Pros and Cons

Pros	Cons
A preliminary idea about the lipid can be obtained	This test is not specific

Alternative Methods/Procedures

Solubility test, grease spot test.

DETERMINATION OF DEGREE OF UNSATURATION OF FATTY ACIDS
Rationale

Unsaturated fats are mostly obtained from plants, while animals are the common source of saturated fats. The unsaturated fats contain a double or triple bond in the hydrocarbon chain of their fatty acids. The halogens such as iodine can be accommodated in these unsaturated bonds. In this test, the Huble's reagent comprising iodine is added to the lipid and the degree of the unsaturation can be determined by the decolorizing of the reagent.

Materials, Equipment, and Reagents

A. **Reagents**: Test samples, Huble's reagent, chloroform.
B. **Glassware**: Test tubes, dropper.

Protocols

1. Take 3 mL of chloroform in the test tubes and label them.
2. Now add 3 mL of Huble's reagent to each test tube.
3. Shake well.
4. The solution turns pink because of the presence of free iodine.
5. Now add the test samples in a dropwise manner.
6. Note down the number of drops required for the pink color to disappear from the solution.

Precursor Techniques

1. Huble's reagent: (a mixture of 7% mercury chloride in alcohol and 5% iodine in 96% alcohol is added in equal proportion). Add 3.5 mL of mercury chloride ($HgCl_2$) to a final solution of 50 mL in ethanol. In a separate beaker, add 2.5 g of iodine in 48 mL of ethanol to a final volume of 50 mL. Now mix both the solutions in equal ratio to a final volume of 100 mL.

Analysis and Statistics

The number of drops of lipid samples required for the disappearance of pink color is more for a lipid having an unsaturated fatty acid with less number of unsaturated bond than that having more number of saturated bonds. In other words, the more the unsaturation the less the number of lipid drops required.

Safety Considerations and Standards

1. Glassware should be washed before and after use.
2. Carefully note down the mL required in the process.
3. The change in the color of the solution should be monitored carefully.

Pros and Cons

Pros	Cons
The degree of unsaturation of fats can be easily determined	The exact number of double bonds cannot be predicted

Summary

1. The degree of unsaturation is indicated by the amount of iodine solution accommodated in a lipid solution.
2. The pink color of the Huble's reagent with chloroform disappears with the addition of fatty acids to it.
3. The more the number of drops of fatty acids required, the less is the degree of unsaturation of fatty acid.

DETERMINATION OF THE SAPONIFICATION VALUE OF THE GIVEN FAT/OIL SAMPLE
Rationale

Saponification is a process involving the hydrolysis of fats on its reaction with alkali, thereby leading to the formation of salts of fatty acids and glycerol. The salts of fatty acids are known as soap. The amount of potassium hydroxide required in the hydrolysis indicates the saponification value of fat. It can be simply described as the amount of alkali (e.g., potassium hydroxide) required to saponify 1 g of fat. Each molecule of triacylglycerol requires three molecules of KOH for saponification. Also 1 g of triacylglycerol with shorter length of fatty acids has more number of molecules of triglycerides as compared to the same with a longer length of fatty acids; therefore, the former will require more amount of KOH. In this way, the saponification value depicts the average molecular weight of a triglyceride. In this process, excess of alkali is used and the remaining KOH is determined by titrating it with 0.5 N HCl.

$$
\begin{array}{llll}
CH_2OR & & CH_2OH & \\
| & & | & \\
CHOR & + 3KOH \rightarrow & CHOH & + 3RCOO^-K^+ \\
| & & | & \\
CH_2OR & & CH_2OH & \\
\text{Fat} & \text{Alkali} & \text{Glycerol} & \text{Soap}
\end{array}
$$

Materials, Equipment, and Reagents

A. **Reagents**: Fat solvent (mixture of 95% ethanol and ether in 1:1 v/v), 0.5 N alcoholic KOH, 1% phenolphthalein solution in 95% alcohol, 0.5 N HCl, test sample.
B. **Glassware**: Burette, conical flask.
C. **Instruments**: Water bath, reflux condenser.

Protocol

1. Weigh 1 g of fat sample in a conical flask and dissolve it in 3 mL of the solvent of fat.
2. To this add 25 mL of 0.5 N of alcoholic KOH, attach a reflux condenser to it.
3. Parallelly, set up another reflux condenser with 3 mL fat solvent and 25 mL of 0.5 N of alcoholic KOH as a blank as it is devoid of fat.
4. Heat both condensers on boiling water bath for 30 min.
5. Let both the flasks to cool to the room temperature.
6. Add a few drops of phenolphthalein to both the flasks.
7. Titrate both with 0.5 N HCl, till pink color disappears.
8. The difference between the blank and test sample gives the volume of KOH required to saponify 1 g of fat.

Precursor Techniques

1. Fat solvent: Mix 47.5 mL of ethanol and 2.5 mL of water to prepare 95% ethanol, mix 95 mL ethanol and 5 mL water. Take 25 ml of 95% ethanol and mix 25 mL ether in it.
2. 1% phenolphthalein solution in 95% alcohol: Mix 95 mL of ethanol and 5 mL of water, to this add 1 g of phenolphthalein.
3. 0.5 N HCl: Add 4.44 mL of conc. HCl to water to make a final solution of 100 mL.

Calculation

Suppose,
0.5 N KOH in the test sample is X mL.
0.5 N KOH in blank is Y mL.

Then, the titer value for the sample is $(Y - X)$ mL = mL of KOH required to saponify 1 g of fat.

It is also known as the saponification value of the fat.

$$
\text{Saponification value}_{\text{(mg/g)}} = \frac{28.05 \times \text{titre value in mL}}{\text{Weight of sample in gram}}
$$

Since 1 mL of 0.5 N KOH contains 28.05 mg of KOH, the multiplication factor of 28.05 is included in the above equation.

The molecular weight of KOH is 56 and a triglyceride releases three molecules of fatty acids.

Therefore,

$$
\text{Saponification value}_{\text{(mg/g)}} = \frac{3 \times 56 \times 1000}{\text{Average molecular weight of fat}}
$$

OR

$$
\text{Average molecular weight of fat} = \frac{3 \times 56 \times 1000}{\text{Saponification value (mg/g)}}
$$

Analysis and Statistics

The saponification value in mg/g and the average molecular weight of fat in g can be calculated.

Safety Considerations and Standards

1. Alcohol is highly inflammable so utmost precaution should be applied while heating.
2. There should be effective cooling of the condenser to prevent the evaporation of alcohol.

Pros and Cons

Pros	Cons
The average molecular weight of fatty acids can be determined	

Summary

1. It is a method to determine the average molecular weight of the fatty acids.
2. Fat is added to a fixed amount of alkali and attached to a reflux condenser. The amount of alkali required to saponify 1 g of fat is determined by titrating it against HCl.

DETERMINATION OF THE FATTY ACID VALUE OF FAT

Rationale

The fats become rancid on storage. The rancidity is caused due to the formation of peroxide by atmospheric oxygen at the double bonds and also because of microorganisms as they cause its hydrolysis to liberate fatty acids and glycerol. Therefore, the age and quality of fats can be determined by the acid value of fat or the amount of free fatty acids present in it. Their presence is determined by the titration of the sample with KOH. The acid value is defined as milligrams of KOH required to neutralize the free fatty acids in 1 g of fat or oil.

Materials, Equipment, and Reagents

A. **Reagents**: 1% phenolphthalein solution in 95% alcohol, fat solvent (95% ethanol: ether in 1:1 v/v), 0.1 N potassium hydroxide.
B. **Glassware**: Burettes, conical flask.

Protocols

1. Take 10 g of fat in a conical flask and add 50 mL of fat solvent to it.
2. Shake the above and add a few drops of phenolphthalein to it.
3. Titrate it with KOH till a faint pink color solution appears for 15 s (approximately).
4. Note down the volume of KOH used.
5. Repeat the above for a blank solution that is devoid of fat.
6. Again, note down the volume of KOH used.

Precursor Techniques

1. Fat solvent: Mix 47.5 mL of ethanol and 2.5 mL of water to prepare 95% ethanol, mix 95 mL ethanol and 5 mL water. Take 25 mL of 95% ethanol and mix 25 mL ether in it..
2. 1% phenolphthalein solution in 95% alcohol: Mix 95 mL of ethanol and 5 mL of water, to this add 1 g of phenolphthalein.
3. 0.1 N potassium hydroxide: Add 2.8 g of KOH in 500 mL of solution in water.

Calculation

The volume of KOH used for blank = X mL.
The volume of KOH used for sample = Y mL.
Therefore, the titer value of fat = $Y - X$ mL.

$$\text{Acid value}_{(mg/g)} = \frac{5.6 \times \text{titre value in mL}}{\text{Weight of sample in gram}}$$

Since, 1 mL of 0.1 N KOH contains 5.6 mg of KOH, the multiplication factor of 5.6 is included in the above equation.

Safety Considerations and Standards

1. Glassware should be washed before and after use.
2. Carefully note down the mL required in the process of titration.
3. The change in the color of the solution should be monitored carefully.

Pros and Cons

Pros	Cons
The quality of fat can be easily determined	

Summary

1. The fatty acid value of fat is determined by titrating it against an alkali.
2. It depicts the age and quality of the fat.

ACROLEIN TEST FOR THE PRESENCE OF GLYCEROL

Rationale

When glycerol is heated with potassium hydrogen sulfate, it gets dehydrated to acrolein, which is an unsaturated aldehyde. This glycerol can be in its free form or as alcohol in fats. Acrolein has a characteristic pungent smell.

$$\begin{array}{lcl} CH_2OH & & CHO \\ | & & | \\ CHOH & + KHSO_4 \xrightarrow{heat} & CH & + 2H_2O \\ | & & || \\ CH_2OH & & CH_2 \\ Glycerol & & Acrolein \end{array}$$

Materials, Equipment, and Reagents

A. **Reagents**: Test samples, anhydrous potassium hydrogen sulfate.
B. **Glassware**: Test tubes, test tube holders.
C. **Instrument**: Bunsen burner.

Protocol

1. Take ~1.5 g of potassium hydrogen sulfate in the test tube.
2. Add five drops of sample to it, in case of solid sample take the equivalent weight of the sample.
3. Cover the sample by adding a few more crystals of potassium hydrogen sulfate to it.
4. Heat the test tube on a burner and note the odor of the fumes coming out of the tube.

Precursor Techniques

None

Analysis and Statistics

The pungent smell of the fumes indicates the presence of glycerol in the test sample.

Safety Considerations and Standards

1. Be careful while heating. Do not inhale the fumes.

Alternative Methods/Procedures

Dichromate test for glycerol.

Summary

1. The test is to detect the glycerol in a sample.
2. The pungent smell on heating the sample with potassium hydrogen sulfate is the indicator of the glycerol.

LIBERMANN-BUCHARD TEST FOR THE DETECTION OF CHOLESTEROL

Rationale

Libermann-Buchard test is a colorimetric test for cholesterol, which yields a deep green color. Although the exact nature of chromophore is not known, it is suggested that in this the hydroxyl group of cholesterol reacts with acetic anhydride and conc. sulfuric acid, this increases the conjugation of the unsaturation of the adjacent fused ring.

Materials, Equipment, and Reagents

A. **Reagents**: Stock solution, working solution, chloroform, Libermann-Buchard reagent.
B. **Glassware**: Test tubes, test tube holders.
C. **Instrument**: Spectrophotometer.

Protocol

1. Take six test tubes and label them.
2. To these add 0–2 mL of working standard solution and add chloroform to make the final volume to 2 mL.
3. Simultaneously, take 2 mL of the sample solution in another test tube.
4. Add 2 mL of Libermann-Buchard reagent, mix well, and keep in dark for 90 min.
5. Measure the optical density at 680 nm.
6. Subtract the value of the absorbance of blank from the absorbance value for each of the test tubes.
7. Plot the standard calibration curve with concentration plotted on the X-axis and optical density on the Y-axis.

Test Tube No.	Blank	1	2	3	4	Test Sample
Working solution (mL)	0	0.5	1.0	1.5	2	2.0
Chloroform added (mL)	2	1.5	1.0	0.5	0	0
Liberman-Buchard reagent	2	2	2	2	2	2
OD at 680 nm	A_0	A_1	A_2	A_3	A_4	A_T
Final OD	A_0	A_1-A_0	A_2-A_0	A_3-A_0	A_4-A_0	A_T-A_0

Precursor Techniques

1. Libermann-Buchard reagent: 50 mL of acetic anhydride is pipetted in a glass vial, keep it in ice. Also 5 mL of concentrated sulfuric acid is added to the vial.
2. **Cholesterol standard:**
   Stock standard (2 mg/mL of cholesterol): Weigh 200 mg of cholesterol and add 100 mL of chloroform to it.
   Working standard (0.4 mg/mL of cholesterol): Dilute 2 mL of stock solution to a final volume of 10 mL by adding chloroform to it.

Analysis and Statistics

The concentration of cholesterol is obtained by spectrophotometer.

Safety Considerations and Standards

1. Be careful while heating. Do not inhale the fumes.

Pros and Cons

Pros	Cons
The widely accepted method for the determination of cholesterol in the blood	The reagents like acetic anhydride, sulfuric acid, and chloroform are corrosive and toxic

Alternative Method

Salkowski method.

Summary

1. It is a quantitative test for cholesterol.
2. The reaction between the Libermann-Buchard reagent and the sample yields a colored product that can be estimated using the standard graph.

EXTRACTION OF LIPID FROM LEAVES (BLIGH AND DYER'S METHOD AND SINGLE-EXTRACTION METHOD)

Rationale

One phase system is used in Bligh and Dyer's method, with chloroform:methanol:water (1:2:0.8), with the tissue water included in the water amount. The extraction is followed by the addition of chloroform and methanol to make two phases. Lipid is found in the chloroform phase. BHT acts as an antioxidant. KCl form cations, which aid in shifting of lipid separation to the organic phase.

Materials, Equipment, and Reagents

A. **Reagents**: Test samples, anhydrous potassium hydrogen sulfate, 0.01% butylated hydroxytoluene (BHT).
B. **Glassware**: Test tubes, test tube holders.
C. **Instrument**: Bunsen burner.

Protocol

Bligh and Dyer's method

1. Heat 4 mL isopropanol with 0.01% BHT at 75°C in a 50 mL tube.
2. Harvest three leaves of Arabidopsis and add it to the tube.
3. Heat the samples at 75°C for 15 min.
4. Let the samples cool down to RT.
5. Add 1.5 mL of chloroform and 0.6 mL of water to the samples.
6. Shake the samples for 1 h.
7. Transfer the extracts to a new tube and extract again the leaf materials four more times with 4 mL chloroform:methanol (2:1) with 0.01% BHT with 30 min of shaking.
8. Combine all the extracts.
9. Wash the extracts with 1 mL of 1 M KCl and then with 2 mL of water.
10. Evaporate the combined extract and dissolve it again in 1 mL chloroform.
11. Transfer the intact, extracted leaf materials to a new vial using forceps and weigh it.

Single-step extraction method

1. Heat 4 mL isopropanol with 0.01% BHT at 75°C in a 20 mL tube.
2. Harvest three leaves of Arabidopsis and add it to the tube.
3. Heat the samples at 75°C for 15 min.
4. Let the samples cool down to RT.
5. Add 12 mL of a mixture of chloroform:methanol:water in a ratio of 30:41.5:3.5 so that the final ratio of the components is 30:41.5:3.5:25 for chloroform:methanol:water:isopropanol.
6. Shake the extracts at 100 rpm for 24 h in an orbital shaker.
7. Transfer the intact, extracted leaf materials to a new vial using forceps.
8. Let it dry overnight at 105°C.
9. Add 280 μL of 600 mM ammonium acetate to the sample before lipid analysis.

Precursor Techniques

1. 0.01% BHT: Weigh 1 mg of BHT and add it to 10 mL of water.
2. 1 M KCl: Add 7.45 g of KCl to make a 100-mL solution with water.

Analysis and Statistics

The pungent smell of the fumes indicates the presence of glycerol in the test sample.

Safety Considerations and Standards

1. Chloroform is toxic.

Pros and Cons

Pros	Cons
One of the most practiced methods for extraction of a broad range of lipids	Less efficient as compared to other methods of extraction for specific types of lipids
	Usage of chloroform is inappropriate for large-scale production, due to its toxic and carcinogenic nature

Alternative Methods/Procedures

Floch's method.

Summary

1. The method utilizes the phase separation method to isolate the lipids from the samples. Lipids get separate out with the chloroform.

EXTRACTION OF LIPID FROM EGG YOLK AND ESTIMATION OF PHOSPHOROUS CONTENT IN IT

Rationale

The extraction of lipid content from the yolk is desirable due to its high nutritional value. The lipids are extracted based on different solubilities of neutral lipids and polar lipids in various organic solvents. The experiment also includes β-CD. The removal of cholesterol by adsorption by β-CD is a better protocol as it is nontoxic, non-hygroscopic, and chemically stable.

Materials, Equipment, and Reagents

A. **Reagents**: Eggs, β-CD solution 95% ethanol, H_2O_2, 5 M H_2SO_4, KH_2PO_4, and 0.2 mL of 5% $(NH_4)_2MoO_4$.

B. **Glassware**: Test tubes, test tube holders, round bottom flask, Kjeldahl flask.

C. **Instruments**: Homothermic reactor with a magnetic stirrer, centrifuge.

Protocol

1. Crack the egg and place it in a homothermic reactor with a magnetic stirrer.
2. Add 95% ethanol to it.
3. Stir it for 1 h at 65°C.
4. Filter the ethanol extracted fraction and collect it in a beaker.
5. Repeat the extraction process twice.
6. Collect all the extracts in a beaker. The solid egg residue is egg yolk protein free from lipids.
7. Crystalize the extracted fraction at 4°C for 8 h.
8. Centrifuge the above at 2800 g, 10 min at 4°C. Remove the solidified triacylglycerol.
9. Collect the supernatant (lipid fraction devoid of triacylglycerol) and place it in a homothermic reactor with a magnetic stirrer.
10. Add an aqueous solution of β-CD dropwise.
11. The molar ratio of β-CD: cholesterol is set to be 5:1.
12. Incubate it for at 30 min.
13. Centrifuge the above at 2800 g for10 min.
14. The precipitate is cholesterol.
15. After cholesterol removal, transfer the supernatant to a previously weighed round bottom flask.
16. Remove ethanol by rotatory evaporation.
17. Collect the dried phospholipids.

Determination of phosphate present in the extracted lipid

1. Dry the lipid solution on a water bath.
2. Take 40 mg of phosphorylated and 100 mg of nonphosphorylated lipids.
3. Digest both in a micro-Kjeldahl flask with 1 and 2 mL of conc. H_2SO_4.
4. Add H_2O_2 in drops at interval till the mixture becomes colorless, follow it with incubation for 30 min for the decomposition of H_2O_2.
5. Transfer the solution to a volumetric flask and make up to the volume.
6. Now label 11 series of the test tube.
7. Make a standard phosphate solution with 10 μg P/mL.
8. Take 1, 2, 3, and 4 mL of the standard solution in the tubes labeled as 1, 2, 3, and 4, respectively.
9. Take 1, 2, and 3 mL of phosphate lipid obtained above in the test tube nos. 5, 6, and 7, respectively, and 1, 2, and 3 mL of nonphosphorylated lipid in test tube nos. 8, 9, and 10, respectively.
10. Take 4 mL of water in the test tube no. 11, this will serve as blank.
11. Make up the volume of test tubes to 4 mL by adding water.
12. Add 5 mL of 5 M H_2SO_4 and 0.2 mL of 5% $(NH_4)_2MoO_4$ solution.
13. To this add 0.3 mL of reducing agent.
14. Shake all the tubes vigorously and heat for 10 min in a boiling water bath.
15. Cool down the tubes and make the volume up to 50 mL with distilled water.
16. Read the absorbance at 630 nm with reference to the blank.

Precursor Techniques

1. 1% NaCl: Add 1 g of NaCl to a final volume of 100 mL in water.
2. Standard phosphate solution: Add 4.392 g of KH_2PO_4 in 1 L water. Take 10 mL of it and make final volume 1 L by water.
3. 5% $(NH_4)_2MoO_4$: Add 5 g of $(NH_4)_2MoO_4$ to a final volume of 100 mL.
4. Reducing agent: Dissolve 3.75 g of sodium metabisulfite $Na_2S_2O_5$, 125 mg of Na_2SO_3, and 65 mg of 1-amino-2-naphthol-4-sulfonic acid in 25 mL water.
5. 95% ethanol: Add 95 mL of ethanol to make a final volume of 100 mL with water.
6. β-CD solution: Add 15 mL of water to per gram of β-CD.

Analysis and Statistics

The phospholipid and nonphosphorylated lipids were extracted from the egg yolk. The determination of phosphorous content in phospholipid and nonphospholipids was done.

Safety Considerations and Standards

1. Handle the egg yolk properly.
2. Care should be taken while working with H_2SO_4.

Summary

1. The extraction of phospholipids depends on the fact that it is more polar in nature.
2. The triglyceride is separated from the mixture by solidifying it by cooling and cholesterol is removed using the β-CD solution.

EXTRACTION OF LIPID FROM TISSUES (FOLCH METHOD)

Rationale

This protocol includes a step of homogenization of a tissue sample in a mixture of chloroform and methanol. Methanol serves as a polar component, which increases the solubilization of lipids present in the cells. Water and methanol, being polar, comprises salt while the chloroform phase contains lipids.

Materials, Equipment, and Reagents

A. **Reagents**: Tissue, chloroform-methanol mix, distilled water, pure solvent upper phase.
B. **Glassware**: Test tubes, glass-stoppered vessel.
C. **Instrument**: Centrifuge.

Protocol

1. Weigh 1 g of tissue and homogenize it with 20 mL of 2:1 chloroform-methanol mixture.
2. Filter the homogenate in a glass-stoppered vessel.
3. Mix the crude extract with 0.2 of its volume of water (i.e., 4 mL).
4. Let it stand (or centrifuge it at 2400 rpm for 20 min) to separate in two phases (the volume of upper and lower phases are 40% and 60%, respectively).
5. Remove the upper phase, as much as possible, by siphoning.
6. Add 1.5 mL of pure solvent upper phase gently, rotate the tube gently to allow the mixing of rinsing fluid with remaining upper phase. Now remove it.
7. Rinse like above twice.
8. Take out the upper phase.
9. The lower phase consists of lipids.

Precursor Techniques

1. Chloroform-methanol mixture: Mix 20 mL of chloroform and 10 mL of methanol.
2. Pure solvent upper phase: Mix chloroform, methanol, and water in a separatory funnel in 8:4:3 by volume. Let it stand, this will result in a biphasic system. The upper part of it is a pure solvent upper phase, while the lower is a pure solvent lower phase. In the upper phase, the proportion of chloroform, methanol, and water is 3:48:47 and in the lower phase, it is 86:14:1.

Analysis and Statistics

The lower phase containing the lipid portion of tissue is obtained.

Safety Considerations and Standards

1. Handle the upper and lower phase carefully, do not mix it.

Pros and Cons

Pros	Cons
The most suitable method for the extraction of a broad range of lipids	Less efficient as compared to the hexane-isopropanol method) for nonpolar lipids

Alternative Methods/Procedures

Bligh and Dyer's method.

Summary

1. The lipid, being nonpolar in nature, is extracted with a nonpolar phase of the extraction solution.

CHAPTER 4

Protein

A. Isolation (extraction) of protein
  1. Isolation of protein from animal cells/tissue.
B. Qualitative test for proteins and amino acids
  1. Biuret test.
  2. Xanthoproteic test.
  3. Millions test.
C. Quantitative test (estimation) for proteins
  1. Biuret test.
  2. Bradford method.
  3. Bicinchoninic acid (BCA) method.
  4. Folin-Lowry method.
  5. UV method.
D. Separation of proteins/amino acid
  1. Protein separation by gel electrophoresis under denaturing condition (SDS-PAGE).
  2. Separation of amino acids by paper chromatography.

ISOLATION OF PROTEINS FROM ANIMAL CELLS/TISSUE (USING RADIOIMMUNOPRECIPITATION ASSAY BUFFER)

Definition

Protein extraction or isolation from tissues/cultured cells or any other sample is the first step for any biochemical as well as molecular analysis (protein purification, PAGE, western blotting, crystallization, proteomics, etc.). For protein extraction, cells are first lysed using appropriate lysis buffer, then proteins are isolated. Different lysis buffers are used for extracting proteins from the tissues or cultured cells depending upon the cellular fractions (location) needed for proteins. Radioimmunoprecipitation Assay (RIPA) buffer is used for extracting whole-cell proteins as well as membrane-bound and nuclear fraction.

Rationale

RIPA lysis buffer is used for rapid lysis of cells and efficient solubilization of proteins from membranes, nuclear as well as cytoplasmic fractions. The RIPA buffer contains ionic [sodium dodecyl sulfate (SDS)] as well as nonionic (Triton X-100) detergents that disrupt the membrane and separate the membrane-bound, cytoplasmic proteins in the buffer.

Materials, equipment, and reagents

A. **Reagents**: Tris base, HCl, NaCl, EDTA, NP-40, sodium deoxycholate, SDS, KCl, Na_2HPO_4, KH_2PO_4.
B. **Glassware**: Microcentrifuge tubes, pipette, tips, cell scraper.
C. **Instruments**: Tissue homogenizer, refrigerated microcentrifuge.

Protocols

A. **From animal tissue**:
  1. Dissect the animal, remove the required organ, and cut into small pieces.
  2. Wash the tissue in PBS buffer and transfer it to a microcentrifuge tube. (Store in liquid nitrogen or −80°C for later use.)
  3. Use 1 mL of ice-cold RIPA buffer for 100 mg tissue (this volume of tissue weigh can vary).
  4. Homogenize the tissue with the help of a homogenizer (with an interval of 30 s).
  5. After homogenization keep the tube in ice for 10 min.
  6. Centrifuge the tube at 13,000 rpm in a refrigerated microcentrifuge for 15 min.
  7. Transfer the supernatant layer to a separate tube.
  8. Aliquot and store at −20°C or −80°C.
B. **From cultured cells**:
  1. Aspirate the culture media form the adherent cells.
  2. Wash the cells twice with ice-cold PBS buffer.
  3. Scrap the cells in 1 mL PBS buffer (in 100 mm dish) with help of scrapper.
  4. Centrifuge to pellet down the cells at 5000 rpm for 10 min at 4°C.
  5. Discard the supernatant PBS buffer and wash the pellet again in ice-cold PBS (optional).
  6. Add 1 mL of ice-cold RIPA buffer and lyse the cells.
  7. Tap the tube for some time and keep in ice for 5–10 min. Repeat this procedure for 30–40 min till the cell pellet lysed completely.
  8. Centrifuge the tube for 15 min at 13,000 rpm at 4°C.
  9. Carefully transfer the supernatant in the separate tube and discard the debris at the bottom.

Precursor techniques

1. Phosphate buffer saline (PBS): Dissolve 8 g NaCl, 0.2 g KCl, 1.44 g Na_2HPO_4, and 0.24 g KH_2PO_4 in

800 mL distilled water. Adjust the pH 7.4 with HCl. Make the final volume of 1000 mL.

2. RIPA buffer:

Reagents	Stock	Working	Volume for 10 mL Buffer
Tris-HCl pH 7.4	1 M	50 mM	500 μL
NaCl	4 M	150 mM	375 μL
EDTA (pH 8)	0.5 M	1 mM	20 μL
SDS	10%	0.1%	100 μL
Triton X-100	100%	1%	100 μL
Na-deoxycholate			10 mg
Distilled water			Remaining to make the volume 10 mL
Protease inhibitors			3–4 μL
Phosphatase inhibitors or sodium orthovanadate			3–4 μL

- 1 M Tris HCl (pH 7.4): Dissolve 121.1 g Tris base in 800 mL distilled water. Adjust the pH 7.4 using HCl. Make the final volume of 1000 mL and autoclave.
- 4 M NaCl: Dissolve 58.44 g NaCl in 250 mL of distilled water.
- 0.5 M EDTA (pH 8): Dissolve 18.61 g disodium ethylenediaminetetraacetate. 2H$_2$O into 80 mL of distilled water. Mix and shake vigorously using a magnetic stirrer. Adjust the pH 8 using NaOH pellets. Once the pH adjusted and salt dissolves, make the final volume 100 mL by distilled water and autoclave.
- 10% SDS: Dissolve 10 g SDS in 100 mL distilled water.
- Amount of RIPA buffer used for cultured cells (grown in).

100 mm dish	1 mL
60 mm dish	0.5–0.6 mL
35 mm dish or 6 well plate	0.3 mL
12 well plate	0.2 mL
24 well plate	0.1 mL

Safety considerations and standards

1. Protease and phosphatase inhibitors must be added just before use.
2. All the steps for cell lysis and protein extraction must be carried out at 4°C or ice.
3. Autoclave all the reagents and solutions after preparations.

Analysis and statistics

The isolated proteins are quantitated by an appropriate method and stored at −20°C.

Pros and cons

Pros	Cons
Extracted proteins are compatible with various assays	Not suitable for study protein-protein interaction
The lysis buffer does not interfere with the immunoreactivity hence suitable for western assay, immunoassays	

Alternative methods/procedures

NP-40 buffer, Laemmli sample buffer (LSB).

Troubleshooting and optimization

Problem	Solution
Low protein concentration	Use less RIPA buffer
Protein degradation	Use protease inhibitors in the buffer just before use
Low protein yield	Lyse the cells completely in the buffer and incubate in the buffer for a longer time

Summary

1. The cultured cells are first harvested, then lysed in RIPA buffer. The lysed cells are centrifuged at 13,000 rpm for 15 min at 4°C. The solubilized proteins in the supernatant is then transferred to another tube.
2. The animal tissue is dissected and cut into pieces. The tissues are homogenized in RIPA buffer, centrifuged, and the supernatant is transferred to another tube.

Definition

To detect the presence of protein in the given sample by the Biuret test.

Rationale

Biuret test is used for the detection of peptide bonds in the protein. This test is not suitable for amino acid only. The protein solution when reacts with CuSO$_4$ in alkali condition (NaOH/KOH), purple to violet-colored copper (II) cation complex is formed. The intensity of the complex is directly proportional to the number of the peptide bond in the protein sample.

Protein + CuSO$_4$ $\xrightarrow{\text{NaOH}}$ Copper cation complex (purple to violet)

Materials, equipment, and reagents

A. **Reagents**: Sodium hydroxide, copper sulfate, sample.
B. **Glassware**: Test tube, test tube holder, dropper, beaker, spirit lamp.

Protocols

1. Take 2 mL of samples in different test tubes.
2. Add 2 mL of 40% NaOH solution in each of the test tubes.
3. Add four to five drops of 1% $CuSO_4$ in the solution.
4. Warm the mixture for about 5 min (optional).
5. Observe the color change.

Precursor techniques

1. 40% NaOH—dissolve 40 g NaOH in 100 mL of water.
2. 1% $CuSO_4$—dissolve 1 g $CuSO_4$ in 1 mL of water.

Safety considerations and standards

1. Glassware should be washed before and after use.
2. Handle the chemicals carefully.
3. Keep the test tube away from the body while heating.

Analysis and statistics

Color changes to violet indicate the presence of proteins in the sample. No change in color means the absence of protein.

Pros and cons

Pros	Cons
Very easy and quick assay	Less sensitive
The assay does not depend on the amount and composition of amino acids in the protein	Detection interfered by the presence of other contaminants like lipid, nucleic acids in the sample

Alternative methods/procedures

Millions test, Xanthoproteic test, Ninhydrin test.

Summary

1. This is a widely used method for the identification of proteins.
2. The reaction of the peptide bond in the protein with $CuSO_4$ in alkali conditions forms a purple-violet complex.
3. This method can be used for the detection of proteins in the biological fluid as urine, plasma, blood, etc.

BIURET TEST

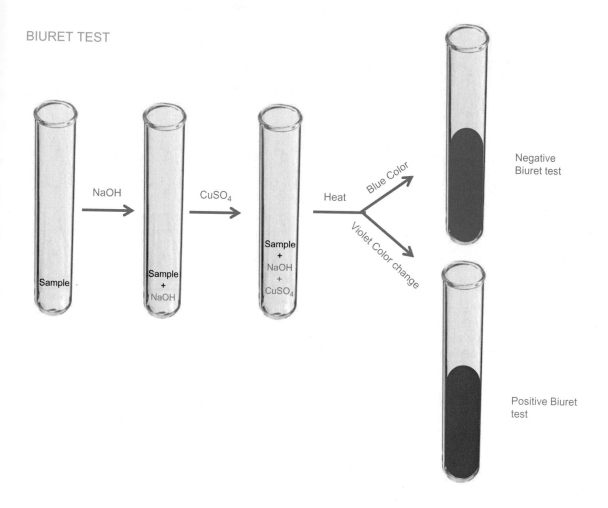

Definition

To detect the presence of protein in the given sample by the Xanthoproteic test.

Rationale

Biuret test is used for the detection of proteins with tyrosine and tryptophan amino acid in their composition. Phenylalanine does not give a positive result.

Proteins having these amino acids give yellow color when heated with conc. HNO_3. The aromatic benzene ring undergoes nitration. Upon adding the alkali solution (NaOH), the yellow color changes to orange. Thus, the appearance of the orange color in a solution with HNO_3 and NaOH indicates the presence of protein with tyrosine and tryptophan amino acids. This test also gives positive results with a solution that contains only tryptophan or tyrosine.

$$\text{Protein} + \text{conc.} HNO_3 \xrightarrow{\text{Heating}} \text{Yellow nitro compound} \xrightarrow{\text{NaOH}} \text{Orange color}$$

Materials, equipment, and reagents

A. **Reagents**: Sodium hydroxide, conc. HNO_3, sample, tyrosine, tryptophan powder.
B. **Glassware**: Test tube, test tube holder, dropper, beaker, spirit lamp.

Protocols

1. Take 1 mL of the sample in a test tube.
2. Take 1 mL of distilled water in another test tube as a negative control.
3. Take 1 mL of 1% tryptophan or 1% tyrosine solution in another test tube as a positive control (optional).
4. Add 1 mL of conc. HNO_3 in each of the test tubes.

5. Mix well and heat for some time.
6. Cool and add 1 mL of 40% NaOH from the side of the test tube.
7. Observe the color change.

Precursor techniques
1. 40% NaOH—dissolve 40 g NaOH in 100 mL of water.

Safety considerations and standards
1. Handle the acid very carefully.
2. Keep the test tube away from the body while heating.

Analysis and statistics
Color changes to orange indicate the presence of proteins in the sample. No change in color means the absence of protein. In the test tube with distilled water, no color occurs and in the test tube with tyrosine or tryptophan orange color appears.

Pros and cons

Pros	Cons
Very easy	Use of strong acid
Can be used to differentiate tyrosine and tryptophan amino acid solution from others	

Alternative methods/procedures
Millions test, Biuret test, Ninhydrin test.

Summary
1. This test is used for the detection of proteins that consist of certain aromatic amino acids like tyrosine and tryptophan.
2. In the detection of the amino acids, only tyrosine and tryptophan give a positive result.
3. The proteins with tyrosine and tryptophan amino acids when heated with conc. HNO_3 cooled and added NaOH gives orange color. The aromatic benzene erring undergoes nitration.

XANTHOPROTEIC TEST

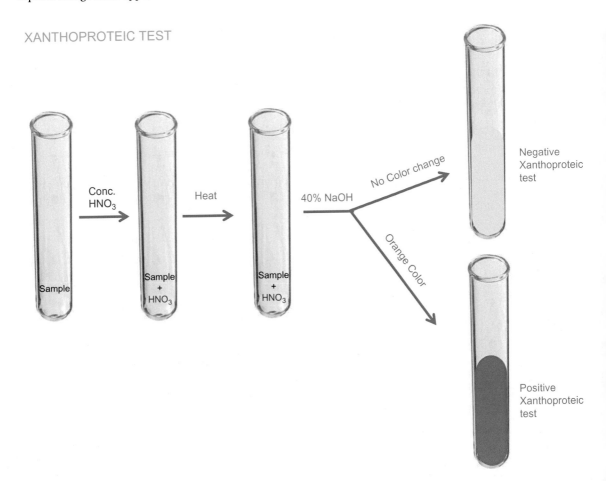

Definition

To detect the presence of protein in the given sample by Million's test.

Rationale

This test is given by phenolic amino acid as tyrosine. Any compounds that have hydroxybenzene group (hydroxyphenyl —C_6H_5OH), forms a red color complex when reacts with Million's reagent. Thus, proteins that contain tyrosine amino acids can be detected by this test. Tyrosine amino acid forms a yellow precipitate of mercury-amino acid complex when treated with acidic mercuric sulfate solution. This solution converts to a red color solution upon reacting with nitric acid or sodium nitrate.

Protein + mercuric sulfate $\xrightarrow{Acid}$ Yellow complex + Sodium nitrite → Red color

Materials, equipment, and reagents

A. **Reagents**: Sodium nitrite, mercuric sulfate, conc. HNO_3, sample, tyrosine, arginine powder.
B. **Glassware**: Test tube, test tube holder, dropper, beaker, spirit lamp.

Protocols

1. Take 1 mL of the sample in a test tube.
2. Take 1 mL of distilled water in another test tube as a negative control.
3. Take 1 mL of 1% tyrosine in another test tube as a positive control and 1% arginine as a negative control (optional).
4. Add 1 mL of the millions reagent in each of the test tubes.
5. Mix well and heat for 1 min.
6. Cool and add four to five drops of 1% $NaNO_2$ in each test tube.
7. Observe the color change.

Precursor techniques

1. Millions reagent: $Hg(SO_4)$ in HNO_3.

Safety considerations and standards

1. Handle the acid very carefully.
2. Keep the test tube away from the body while heating.

Analysis and statistics

Color changes to red indicate the presence of proteins (containing tyrosine amino acid) in the sample. No change in color means the absence of protein. In the test tube with distilled water and arginine, no color change and test tube with tyrosine red color appear.

Pros and cons

Pros	Cons
Very easy	Use of strong acid
Can be used to differentiate tyrosine amino acid solution from others	Chlorides interfere. Not suitable for a urine sample

Alternative methods/procedures

Xanthoproteic test, Biuret test, Ninhydrin test.

Summary

1. Any phenolic compound without any group at 3 and 5 positions like tyrosine gives this test positive.
2. Hydroxyphenyl group of tyrosine or phenolic compound reacts with mercury in acidic conditions to form a red complex.

MILLIONS TEST

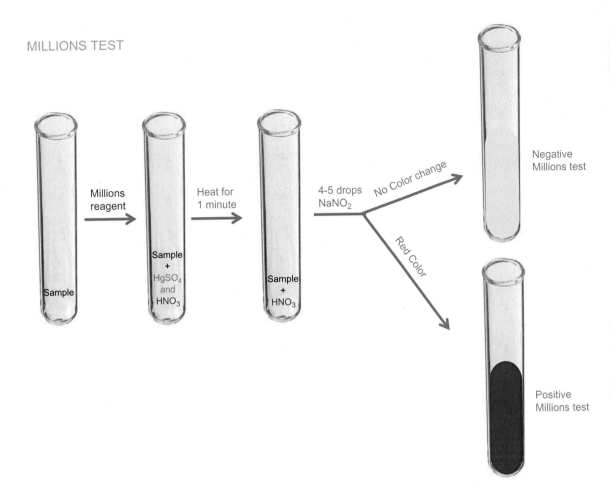

QUANTITATIVE TEST (ESTIMATION) FOR PROTEINS

Proteins are isolated by a suitable method then quantitated for further analysis. Protein estimation is a routine and the most commonly performed tasks in a biochemistry and molecular biology laboratory.

ESTIMATION OF PROTEINS BY THE BIURET METHOD

Definition

The Biuret method is a colorimetric technique to estimate proteins. In this reaction, Biuret reagent is used. This reagent binds with peptide bonds and the colored product intensity is directly proportional to the number of peptide bonds in the sample. This method is not suitable for amino acids.

Rationale

In an alkaline medium, peptide bonds of the proteins combine with cupric ions to form a colored complex (blue-violet). One cupric ion binds with 4–5 peptide bonds. The intensity of the colored complex is directly proportional to the protein concentration in the sample. The absorbance of the colored product can be measured at 530–560 nm (green filter), preferably at 540 nm. At the concentration range of standard protein, there is a linear relationship between the absorbance at 540 and the concentration of the total protein. A standard linear curve can be created to calculate the concentration of an unknown sample. Tartrate (chelating reagent) is added to stabilize the complex in alkaline conditions, while iodide prevents its further reaction.

Reaction:

$$\text{Peptide} + Cu^{2+}(\text{Blue}) \xrightarrow{\text{Alkaline medium}} \text{Peptide} - \text{Copper complex (purple color)}$$

Materials, equipment, and reagents

A. **Reagents**: Cupric sulfate, sodium potassium tartrate (Rochelle salt), potassium iodide, sodium hydroxide, standard protein sample [bovine serum albumin (BSA)], sample.

B. **Glassware**: Test tube, pipette, beaker, measuring cylinder, cuvette, Elisa plate.

C. **Instruments**: Weighing machine, colorimeter/spectrophotometer.

Precursor techniques

1. 10% NaOH: Dissolve 30 g of NaOH in 300 mL distilled water.
2. Biuret reagent: Dissolve 1.5 g cupric sulfate and 6 g of sodium potassium tartrate $(KNaC_4H_4O_6\cdot4H_2O)$ in 500 mL of distilled water. Add 300 mL of 10% NaOH. Make the volume 1 L. Add 1 g of potassium iodide to prevent the reduction of copper. Store in a black bottle in dark.
3. Standard protein solution (10 mg/mL): Dissolve 10 mg of BSA in 1 mL of distilled water. Store it at $-20°C$.

Protocols

A. **One-step method**:

Reagent	Standard	Blank	Test
Biuret reagent	5 mL	5 mL	5 mL
Protein standard (10 mg/mL)	50 μL	–	–
Test sample	–	–	50 μL
Distilled water		50 μL	

Mix thoroughly and keep at room temperature for 20 min
Measure the optical density (absorbance) of the test and standard sample at 540 nm against a blank. Set 100% transmittance using a blank sample

OD_{540}	S		T

B. **Standard curve method**:
1. Preparation of standard curve:
   a. First, prepare a standard series of proteins (1–10 mg) by taking the different volume of protein standard (1 mg/mL) and make the final volume 1 mL as mentioned in the table.
   b. For blank take 1 mL of distilled water.
   c. Add 3 mL of the Biuret reagent in each test tube and incubate all the tubes at room temperature for 30 min.
   d. Set the transmittance of the instrument 100% using blank and read the absorbance at 540 nm in colorimeter or a spectrophotometer.
   e. Plot the standard curve between absorbance and concentration of the standard protein's series.

	Blank	1 mg	2 mg	3 mg	4 mg	5 mg	7 mg	10 mg
Protein standard (10 mg/mL)	–	0.1 mL	0.2 mL	0.3 mL	0.4 mL	0.5 mL	0.7 mL	1 mL
Distilled water	1 mL	0.9 mL	0.8 mL	0.7 mL	0.6 mL	0.5 mL	0.3 mL	–
Biuret reagent	3 mL	3 mL	3 mL	3 mL	3 mL	3 mL	3 mL	3 mL
	Keep at room temperature for 30 min							
Optical density at 540 nm		S1	S2	S3	S4	S5	S6	S7

2. Determination of concentration of an unknown sample: Take 1 mL of an unknown sample, add 3 mL of the Biuret reagent and measure the optical density at 540 nm. Repeat this step thrice with the same sample and take the average of the reading.

Calculation

A. **One-step method**: Concentration of the protein in an unknown sample can be calculated using the following formula:

$$\frac{\text{The absorbance of sample (T)}}{\text{The absorbance of standard (S)}} \times \text{Concentration of standard (10 mg/mL)}$$

B. **Standard curve method**: From the standard curve, the protein present in an unknown sample can be estimated.

Safety considerations and standards

1. Pipetting should be done accurately.
2. The solution before the measurement of absorbance should be mixed well to homogenize the content.

Analysis and statistics

The given sample contains _____ mg/mL of protein.

Pros and cons

Pros	Cons
It can be performed using a colorimeter. No need for a spectrophotometer	Sensitivity is less
Useful for the sample with high protein concentration	Not suitable for the sample where proteins are precipitated using ammonium sulfate
No effect of amino acid compositions	Some buffers as Tris and nucleic acid interfere with the reaction
A fast and easy method	Not high accuracy

Alternative methods/procedures

Bradford method, Lowry method, bicinchoninic acid (BCA) method, etc.

Troubleshooting and optimization

Problem	Solution
Preparation of standard curve: Getting a linear relationship between the concentration of protein and absorbance is sometimes difficult	The pipetting should be done very carefully while making different standard dilutions. The same amount of biuret reagent should be added

Summary

1. The Biuret method is a colorimetric assay to estimate proteins.
2. In this method, peptide bonds in protein react with copper sulfate (Biuret reagent), a blue-violet color is obtained. The intensity of the color depends on the amount of protein or the number of peptide bonds present. The absorbance can be measured at 540 nm. Using a standard plot, the concentration of protein in an unknown sample can be obtained.
3. Sodium potassium tartrate acts as a stabilizer in this reaction.
4. This method is a good method for total protein in the sample as a serum. Total protein concentration increases in a disease like multiple myeloma, liver disease dehydration, etc., while decreases renal disease, liver cirrhosis, etc.
5. Sensitivity: 1–5 mg protein.

ESTIMATION OF PROTEINS BY THE BRADFORD METHOD

Definition

Bradford method is a simple and rapid method of protein estimation. It is based on the binding of Coomassie brilliant blue dye (Bradford reagent) with proteins with the formation of blue color product. The intensity of the color directly related to the amount of protein in the sample. This method is not suitable for amino acids.

Rationale

The protein binds with Bradford reagent, i.e., Coomassie brilliant blue G250 and forms a blue color complex that shows absorption maxima at 595 nm. The intensity of this blue color increases with an increase in protein concentration.

Coomassie blue G dye (Serva blue G) in strong acidic condition presents as the most stable double-protonated red form, which shows absorption maxima at 475 nm. After binding with protein, unprotonated blue form is most stable with absorption at 595 nm. Thus, upon binding with protein, the absorption maxima shift from 475 to 595 nm causing the visible color change. The protein-dye complex stabilizes by both the hydrophobic and ionic bonds.

Materials, equipment, and reagents

A. **Reagents**: Coomassie brilliant blue (CBB) G-250, ethanol, phosphoric acid, standard protein sample [bovine serum albumin (BSA)], sample.
B. **Glassware**: Pipette, beaker, measuring cylinder, flask, Elisa plate, Eppendorf tubes, tips, stand.
C. **Instrument**: Weighing machine, spectrophotometer.

Precursor techniques

1. Bradford reagent: Add 100 mg of CBB G250 in 50 mL ethanol (95%) and dissolve on a shaker for 60 min. Add 100 mL of orthophosphoric acid (85% w/v) and make the final volume of 1000 mL by adding distilled water. Filter through Whatman filter paper 1. Store the solution at 4°C in a dark bottle.
2. Standard protein solution; BSA:
   Stock (1 mg/mL or 1 µg/µL): Dissolve 10 mg of BSA protein in 10 mL of distilled water. Store it at −20°C.
   Working (0.1 µg/µL): Dilute the stock sample by 10 times which means take 100 µL of stock and add 900 µL of distilled water.

Protocols

1. Preparation of standard protein solution: First prepare standard series of proteins (0–5 µg) by taking the different volume of protein standard (0.1 µg/mL) in clean Elisa plate (in triplicate) and add distilled water in well to make final volume 50 µL as mentioned in the table.
2. For the sample, take 5 µL of it and add 45 µL distilled water (in triplicate).
3. Add 200 µL of Bradford reagent in each well.
4. Cover the Elisa plate with aluminium foil and keep it at room temperature for 20 min for the reaction to occur. As the reaction occurs, the red-purple color of the solution turns blue. The intensity of the color developed is directly proportional to the protein concentration thus with increased concentration of protein standards the blue color intensity gradually increases. The color is stable for 20–30 min; after that the protein-dye complex gets precipitate.
5. Read the optical density at 595 nm using a spectrophotometer.
6. Substrate the reading of the blank sample from the standard and test sample reading.

7. Plot a standard curve between standard solution concentration and corresponding absorbance. The standard curve should follow Beer-Lambert law, which means there should be a straight line in the graph plotted between standard protein (ranging 0–5 µg) and their absorbance.

8. From the standard curve, calculate the amount of protein in an unknown sample.

	Blank (B)	0.5 µg (S1)	1 µg (S2)	2 µg (S3)	3 µg (S4)	4 µg (S5)	5 µg (S6)	Test Sample (T)
Protein standard (0.1 µg/µL)	0 µL	5 µL	10 µL	20 µL	30 µL	40 µL	50 µL	5 µL
Distilled water	50 µL	45 µL	40 µL	30 µL	20 µL	10 µL	0 µL	45 µL
Bradford reagent	150 µL	150 µL	150 µL	150 µL	150 µL	150 µL	150 µL	150 µL
	20 min at room temperature							
OD$_{595}$	B	S1-B	S2-B	S3-B	S4-B	S5-B	S6-B	T-B

Calculation

Plot a standard curve between the concentration of the standard protein solution and absorbance at 595 nm. Determine the slope y/x from the standard curve, which gives the A595 per unit of protein (µg). Then determine the protein concentration in an unknown sample.

Safety considerations and standards

1. All the tubes and glassware should be clean, free from reagent, and autoclaved.
2. Weighing of the BSA sample should be done accurately.
3. Pipetting should be done carefully.
4. The dye should be prepared carefully and stored at 4° C. Keep away from light.

Analysis and statistics

The concentration of protein in the given sample _____ µg/µL.

Pros and cons

Pros	Cons
Comparatively less interference. No interference by Triton X-100	Interference by the strong basic buffers, detergent-like SDS, deoxycholate
Very easy, fast and accurate	Not suitable for proteins that are in strong alkaline condition
Compatible with many buffers	Depends strongly on the amino acid composition
Very sensitive and cheap method	Dilution is required before analysis for better sensitivity

Alternative methods/procedures

Lowry method, bicinchoninic acid (BCA) method, Kjeldahl method, etc.

Troubleshooting and optimization

Problem	Solution
The detection limit of low molecular weight protein is 3–5 kDa by this method	Use of another method like BCA for very low molecular weight proteins
Some of the detergents or buffers in the sample interfere the accuracy	Dilution of the sample so that buffer or detergent amount also minimized
The absorbance of the sample is higher than expected due to the high concentration of protein	Dilution of the sample

Summary

1. Bradford method is a routinely used method for protein estimation. This method is sensitive to the detection of a very low amount of protein in the sample. Sensitivity 1–20 µg.
2. This method is based on the binding of CBB G250 dye with protein, which results in the shifts of absorbance of the dye from 475 to 595 nm. With an increasing amount of protein in the sample, the absorbance is also increased at 595 nm.

ESTIMATION OF PROTEINS BY THE BCA METHOD

Definition

This method of protein estimation was first developed by Smith and coworkers in 1985. This method is very simple, sensitive, and suitable for many detergents. The method is based on the binding of bicinchoninic acid (BCA) with cuprous ions produced by the reduction of cupric ion by protein in an alkaline medium.

Rationale

Under the alkaline condition, the cupric ions are reduced to cuprous ions by a protein that reacts with BCA to form a purple-blue complex. This complex shows maximum absorption at 562 nm. The intensity of the color vis-à-vis absorbance is in direct relation to the concentration of protein. Peptide bonds and some amino acids as cysteine, cystine, tryptophan, and tyrosine are capable to reduce Cu^{2+} to Cu^+, thus color complex formation with BCA. Two molecules of BCA form complex with one molecule of Cu^+. The protein concentration in the sample determined with reference to the standard protein as bovine serum albumin (BSA).

Reaction:

$$\text{Protein (peptide bonds)} + Cu^{2+} \longrightarrow Cu^{1+}$$

$$Cu^{1+} + 2 \text{ Bicinchoninic acid} \longrightarrow 2BCA - Cu^{1+}\text{Complex}$$
$$\text{(purple colored Complex}$$
$$\text{Absorbance at 562 nm)}$$

Materials, equipment, and reagents

A. **Reagents**: Bicinchoninic acid, sodium carbonate, sodium bicarbonate, sodium tartrate, sodium hydroxide, copper sulfate, standard protein sample [bovine serum albumin (BSA)], sample.

B. **Glassware**: Pipette, beaker, measuring cylinder, flask, cuvette, Elisa plate, Eppendorf, tips, stand.

C. **Instrument**: Weighing machine, spectrophotometer.

Precursor techniques

1. Sodium hydroxide solution (5 N): Dissolve 20 g NaOH in 100 mL distilled water.

2. Bicinchoninic acid solution (reagent A): Dissolve 1 g BCA, 2 g sodium carbonate, 0.16 g sodium tartrate, 0.4 g NaOH, and 0.95 g sodium bicarbonate in 100 mL with distilled water. Adjust the pH to 11.25 using 5 N NaOH. Store in a brown bottle at room temperature.

3. Copper sulfate solution, 4% (reagent B): Dissolve 4 g $CuSO_4 \cdot 5H_2O$ in 100 mL distilled water.

4. Working BCA reagent (B+C): Just before use mix 50 mL of reagent A with 1 mL of reagent B.

5. Standard protein solution; BSA (1 mg/mL or 1 µg/µL): Dissolve 1 mg of BSA protein in 1 mL of distilled water. Store it at −20°C. Add a small quantity of sodium azide to prevent microbial contamination.

Protocols

1. Preparation of standard protein solution: First prepare standard series of proteins (0–50 µg) by taking the different volume of protein standard (1 µg/µL) in a clean tube (in triplicate) and add distilled water to make the final volume 100 µL as mentioned in the table. Tube with 0 µg of protein (only water) considered blank.

2. For the test sample, take 5 µL of it and add 95 µL distilled water (in triplicate).

3. Add 2 mL of working BCA reagent in each tube.

4. Incubate it at room temperature for 2 h (preferable 30 min at 37°C).

5. Read the absorbance at 562 nm using a spectrophotometer/colorimeter.

6. Substrate the reading of blank samples from the standard and test sample reading [or the first set the transmittance 100% (absorbance 0) with blank sample].

7. From the reading of the standard sample, plot a graph between concentration (horizontal axis) and absorbance (vertical axis). The standard curve should follow Beer-Lambert law, which means there should be a straight line in the graph plotted between standard protein (ranging 0–100 µg) and their absorbance.

8. From the standard curve, calculate the amount of protein in an unknown sample.

	Blank (B)	5 µg (S1)	10 µg (S2)	20 µg (S3)	30 µg (S4)	50 µg (S5)	70 µg (S6)	100 µg (S7)	Test Sample (T)
Protein standard (1 µg/ µL)	0 µL	5 µL	10 µL	20 µL	30 µL	50 µL	70 µL	100 µL	5 µL
Distilled water	100 µL	95 µL	90 µL	80 µL	70 µL	50 µL	30 µL	0	95 µL
Working BCA reagent	2 mL	2 mL	2 mL	2 mL	2 mL	2 mL	2 mL	2 mL	2 mL
				2 h at room temperature					
OD_{562}	B	S1-B	S2-B	S3-B	S4-B	S5-B	S6-B	S7-B	T-B

Important note: The procedure can be performed by the test tube method (using a colorimeter) or microplate method (using a spectrophotometer). The above-described method is the test tube method. In the case of the microplate method, all steps are the same except the final volume of protein standard + distilled water should be 25 µL (instead 100 µL) and add 200 µL of BCA working reagent to be added (instead 2 mL).

Calculation

Plot a standard curve between the concentration of the standard protein solution and absorbance at 562 nm. Determine the slope y/x from the standard curve, which gives the A562 per unit of protein (µg). Then determine the protein concentration in an unknown sample.

Safety considerations and standards

1. All the tubes and glassware should be clean, free from reagent, and autoclaved.

2. Weighing of the BSA sample should be done accurately. BSA should be diluted in 0.9% NaCl or PBS buffer.
3. Pipetting should be done carefully.

Analysis and statistics
The concentration of protein in the given sample _____ µg/µL.

Pros and cons

Pros	Cons
Compatible with many ionic and nonionic detergents hence useful for membrane protein estimation	Incubation time for reaction is long (~2 h)
Useful for a wide range of protein concentrations. Very sensitive	Expensive reagent
Less variation in the reading with different proteins	Interference by carbohydrates, lipids, tryptophan, phenol red, cysteine, glucose, tyrosine, uric acid, iron and hydrogen peroxide, ammonium sulfate, drugs like penicillin, paracetamol
The color complex form is stable	
It can be performed in the Eppendorf tube as well as a microtiter plate	

Alternative methods/procedures
Lowry method, Bradford method, Kjeldahl method, etc.

Troubleshooting and optimization

Problem	Solution
Interference by reducing agents and copper chelators, tris buffer, DTT, etc.	Removal of these reagents by precipitation (trichloroacetic acid; TCA or acetone) and dissolving the protein sample in water or 0.1 N NaOH before analysis
Interference with lipids	Extract the sample with ethanol and ether mixture (2:1 V/V), centrifuge, discard upper organic phase. Then extract again in ether only, centrifuge, discard upper phase. Air-dry the material and precipitate protein in 10% cold TCA, leave it on ice for 10 min, centrifuge, discard the supernatant and dissolve the pellet in 0.1 N NaOH
Interference by polyphenols in plant extract	Polyphenols in pant extract interfere with the process as they also get precipitated out with TCA. Add polyvinyl pyrrolidone (PVP) as 1% in the extraction buffer of protein. PVP will form complex with polyphenols and the precipitate can be separated by centrifugation
No color develops in any tube due to copper chelating agent	Desalt or dilute the sample Double the volume of reagent B in the working BCA solution
Color of sample darker than expected due to high protein or lipid contamination	Dilute the sample or remove lipid as mentioned above
Spectrophotometer or colorimeter does not have 562 nm filter	Measure at 540–590 nm.
All tubes including blank appear dark purple due to reducing agent, thiol-containing agents of catecholamines	Dialyze or dilute the sample

Summary
1. Sensitivity from 0.5 µg/mL to 1.5 mg/mL. More sensitive than the Lowry method and less protein-to-protein variation than the Bradford method.
2. This method is based on the reduction of Cu^{2+} to Cu^+ by protein (this reaction is called Biuret reaction). The Cu^+ produces a purple-blue color complex (absorption maxima at 562 nm) with BCA. The intensity of the color with BCA is in a protein concentration-dependent manner.

ESTIMATION OF PROTEINS BY THE FOLIN-CIOCALTEAU (LOWRY) METHOD

Definition
This method of protein estimation is most widely used in molecular biology experiments and research. Earlier a Folin-Denis reagent was used for the detection and estimation of phenols in the sample. As phenol ring is present in tyrosine amino acid, later Folin-Ciocaltaeu was developed for the protein estimation. This method later named as Lowry method because in the year 1951 Lowry et al. use this principle for the estimation of proteins in the biological samples. Proteins are first precipitated by 5% TCA, pelleted by centrifugation, washed, and dissolve in 0.1 N NaOH.

Rationale

In this method, protein reacts with Folin phenol reagent to produce a blue color. The color development occurs in a two-step reaction. The absorbance of this blue color solution can be measured at 660 nm.

First: Biuret reaction: $CuSO_4$ reacts with the protein in an alkaline medium to form a blue color complex. Sodium potassium tartrate acts as a chelating agent.

Second: In the second reaction, the phenol containing amino acid as tyrosine present in the copper-protein complex reacts with phosphomolybdic acid and phosphotungstic acid of Folin-Ciocalteau reagent, which results in the formation of molybdenum blue. Copper act as a catalyst.

Reaction:

$$\text{Peptide (protein)} + Cu^{2+} \rightarrow \text{Copper} - \text{protein complex (Blue)}$$

$$\text{Copper} - \frac{\text{protein}}{\text{complex}} + \text{Folin reagent} \rightarrow \frac{\text{Reduced Folin}}{\text{reagent (660 nm)}}$$

The intensity of the color of the solution is directly proportional to the number of amino acids in the protein.

Materials, equipment, and reagents

A. **Reagents**: Sodium carbonate, sodium hydroxide, copper sulfate, sodium potassium tartrate, commercial Folin-Ciocalteau reagent, standard protein sample [bovine serum albumin (BSA)], sample.

B. **Glassware**: Pipette, beaker, measuring cylinder, flask, Elisa plate, Eppendorf, tips, stand.

C. **Instrument**: Weighing machine, spectrophotometer.

Precursor techniques

1. wReagent A (2% sodium carbonate): First prepare 0.1 N NaOH by dissolving 0.4 g NaOH in 100 mL of distilled water. Weigh 2 g of sodium carbonate and dissolve in 80 mL 0.1 N NaOH then make the final volume 100 mL.

2. Reagent B (2% $CuSO_4\cdot5H_2O$): Dissolve 2 g copper sulfate in 100 mL distilled water.

3. Reagent C (2% sodium/potassium tartrate): Dissolve 2 g of sodium or potassium tartrate in 100 mL distilled water.

4. Reagent D (freshly prepared): Mix reagent A + reagent B + reagent C in the ratio of 100:1:1.

5. Reagent E (Folin-Ciocalteau reagent): Commercially Folin-Ciocalteau reagent is available at 2 N concentration. Dilute it by distilled water to 1 N before use. Do not use this solution if the 2 N solution shows the absorption of more than 0.2 at 660 nm. This solution is light sensitive.

For manual preparation of the reagent, dissolve 10 g sodium tungstate ($Na_2WO_4\cdot2H_2O$) and 2.5-g sodium molybdate ($NaMO_4\cdot2H_2O$) in 70 mL water. To this, add 5 mL of orthophosphoric acid (85%) and 10 mL of conc. HCl, reflux gently for 10 h. Add 15 g lithium sulfate, 5 mL water. After cooling, add two to three drops of bromine water and mix well. The final solution should be golden yellow and store it in a brown bottle.

6. Standard protein solution; BSA (1 mg/mL or 1 µg/µL): Dissolve 1 mg of BSA protein in 1 mL of distilled water. Store it at −20°C. Add a small quantity of sodium azide to prevent microbial contamination.

Protocols

1. Preparation of standard protein solution: First prepare standard series of proteins (0–100 µg) by taking the different volume of protein standard (1 µg/µL) in a clean tube (in triplicate) and add distilled water to make the final volume 200 µL as mentioned in the table. Tube with 0 µg of protein (only water) considered blank.

2. For the test sample, take increasing amount (as 10, 50, and 100 µL) of it and make final volume 200 µL with distilled water (in triplicate).

3. Add 1 mL of freshly prepared reagent D in each tube.

4. Mix well and incubate it at room temperature for 10 min.

5. Add 0.3 mL of 1 N Folin-Ciocalteau reagent, vortex the tube very fast (within 1–2 s) and incubate the tube at room temperature for 30–60 min.

6. Read the absorbance at 660 or 750 nm using a spectrophotometer/colorimeter.

7. Substrate the reading of blank samples from the standard and test sample reading [or the first set the transmittance 100% (absorbance 0) with blank sample].

8. From the reading of the standard sample, plot a graph between concentration (horizontal axis) and absorbance (vertical axis). The standard curve should follow Beer-Lambert law, which means there should be a straight line in the graph plotted between standard protein (ranging 0–100 µg) and their absorbance.

9. From the standard curve, calculate the amount of protein in an unknown sample.

	Blank (B)	5 µg (S1)	10 µg (S2)	20 µg (S3)	30 µg (S4)	50 µg (S5)	70 µg (S6)	100 µg (S7)	Test Sample (T)
Protein standard (1 µg/µL)	0 µL	5 µL	10 µL	20 µL	30 µL	40 µL	50 µL	100 µL	10, 50, 100 µL
Distilled water	200 µL	195 µL	190 µL	180 µL	170 µL	160 µL	170 µL	100 µL	190, 150, 100 µL
Reagent D (A+B+C)	1 mL	1 mL	1 mL	1 mL	1 mL	1 mL	1 mL	1 mL	1 mL
	10 min at room temperature								
Reagent E (FC reagent)	0.3 mL	0.3 mL	0.3 mL	0.3 mL	0.3 mL	0.3 mL	0.3 mL	0.3 mL	0.3 mL
	Immediate mixing and 30–60 min at room temperature								
OD660	B	S1-B	S2-B	S3-B	S4-B	S5-B	S6-B	S7-B	T-B

Important note: For the microplate method, the basic steps are the same except the volume of different reagents, protein standard, and sample for analysis. The final volume of protein standard + distilled water should be 40 µL (instead 200 µL) and add 200 µL of reagent D (instead 1 mL) and 20 µL of reagent E (instead 0.3 mL).

Calculation

Plot a standard curve between the concentration of the standard protein solution and absorbance at 660 nm. Determine the slope y/x from the standard curve, which gives the A660 per unit of protein (µg). Then determine the protein concentration in an unknown sample.

Safety considerations and standards

1. All the tubes and glassware should be clean, free from reagent, and autoclaved.
2. Weighing of the BSA sample should be done accurately. BSA should be diluted in 0.9% NaCl or PBS buffer.
3. Do not use if the reagent D is turbid.
4. While preparing reagent D, solution B (tartrate) should be mixed well before adding copper sulfate (reagent A) otherwise precipitation of copper may occur.
5. As the half-life of Folin-Ciocalteau reagent is 8 s, hence immediate mixing is required after adding reagent E (Folin-Ciocalteau reagent) in the tube.

Analysis and statistics

The concentration of protein in the given sample _____ µg/µL.

Pros and cons

Pros	Cons
Easy and less expensive	Timing for adding and mixing the reagent needs to be very precise
Highly reproducible	Interference by some buffers and detergents (Tris, MOPS, HEPES, ammonium sulfate, DTT, urea, etc.)
Broad sensitivity range	Sensitivity depends on the amino acid composition of protein
1% SDS is tolerable	Time consuming
	Protein with high proline like gelatin shows less reactivity

Alternative methods/procedures

Bicinchoninic acid (BCA) method, the Bradford method, Kjeldahl method, etc.

Troubleshooting and optimization

Problem	Solution
No color develops in any tube	Dialyze or dilute the sample to minimize interference by chelating agents like EDTA, EGTA
Less color development than expected	Dialyze or dilute the sample to minimize interference by strong acidic or basic conditions
Dark color development than expected (also in the blank) due to a reducing agent or thiol contamination	Dialyze or dilute the sample The interfering substances can be removed by acetone/TCA precipitation followed by centrifugation and dissolving pellets in a suitable buffer
Precipitation due to detergents	Desalt the sample to remove the interfering agents
Instrument does not have 660 nm filter	Measure absorbance at 650–750 nm range

Summary

1. Sensitivity from 5 to 100 µg/mL.
2. This method proceeds in two steps where first is the Biuret reaction where proteins react with the copper sulfate and reduce cupric to cuprous ion. In the second step, the complex reacts with Folin-Ciocalteau reagent, which results in a blue color solution. The absorbance of the compound can be measured at 660/750 nm.
3. The method is also used for the determination of tyrosine and phenolic compounds.

ESTIMATION OF PROTEINS BY UV SPECTROPHOTOMETRIC METHOD

The UV spectrophotometric method is a very simple, rapid, and sensitive method. For the pure protein sample, it is the best method. The electromagnetic spectrum 10–400 nm region is the ultraviolet region. This method is based on the direct measurement of the absorbance of the protein sample at 280 nm. This method was developed by Warburg and Christian in 1941.

Aromatic amino acids (tryptophan and tyrosine) in the proteins show maximum absorption at 280 nm. The absorbance of the protein solution at 280 nm can be a direct determination of protein concentration.

PROTEIN SEPARATION BY GEL ELECTROPHORESIS UNDER DENATURING CONDITION (SDS-PAGE)

Definition

After extraction and quantitation, the next step is the separation of proteins on the gel. In sodium dodecyl sulfate-polyacrylamide gel electrophoresis (SDS-PAGE), proteins are separated in the denature ng condition, where SDS acts as a denaturing agent.

Rationale

In the gel electrophoresis principle, biomolecules are separated on the gel under the influence of the electric charge. The positive ions migrate toward the negative pole and negative ions toward the positive pole. The pore size in the gel restricts the migration thus movements depend upon the size of the molecule.

In SDS-PAGE, SDS denatures the proteins, breaks the disulfide bonds, and binds with its primary structure. The SDS provides the overall negative charge to the molecule and eliminates the influence of the charge and structure on the migration. In SDS-PAGE, the migration solely depends on the size of the molecule thus smaller size proteins travel at a faster rate. Polyacrylamide used for making gel, which provides supporting material for separation.

Materials, equipment, and reagents

A. **Reagents**: Tris base, EDTA, acrylamide, bis-acrylamide, TEMED, ammonium persulfate, SDS, glycerol, β-mercaptoethanol, bromophenol blue, NaCl, prestained protein ladder (optional), Coomassie brilliant blue (CBB), glycine.
B. **Glassware**: Pipette, beaker, measuring cylinder, flask, bottle, plastic, gloves, tips, Eppendorf tube.
C. **Instrument**: Weighing machine, electrophoresis apparatus, autoclave.

Precursor techniques

- 30% acrylamide/bis solution: Dissolve 29 g acrylamide and 1 g N,N-bis-acrylamide (cross-linker) in 100 mL distilled water. Filter it and store it in a black bottle at 40°C.
- 10% APS: Dissolve 0.1 g APS in 1 mL distilled water. Prepare fresh and store at 40°C.
- 1 M Tris HCl (pH 7.4): Dissolve 121.1 g Tris base in 800 mL distilled water. Adjust the pH 7.4 using HCl. pH can be adjusted as per requirements. Make the final volume of 1000 mL and autoclave.
- 1.5 M Tris HCl (pH 8.8): Dissolve 181.65 g Tris base in 800 mL distilled water. Adjust the pH 8.8 using HCl. Make the final volume of 1000 mL and autoclave.
- 4 M NaCl: Dissolve 58.44 g NaCl in 250 mL of distilled water.
- 0.5 M EDTA (pH 8): Dissolve 18.61 g disodium ethylenediaminetetraacetate. $2H_2O$ into 80 mL of distilled water. Mix and shake vigorously using a magnetic stirrer. Adjust the pH 8 using NaOH pellets. Once the pH adjusted and salt dissolves, make the final volume 100 mL by distilled water and autoclave.
- 10% SDS: Dissolve 10 g SDS in 100 mL distilled water.

1. Sample preparation buffer (6× loading dye, Laemmli buffer): In an Eppendorf tube, take 0.6 (6% w/v) g SDS, 6 mg bromophenol blue, 3 mL glycerol, 3 mL from 1 M Tris Cl (pH 6.8), 1 mL of β-mercaptoethanol and make the final volume 10 mL by distilled water.
2. Running buffer (10×): Dissolve 30 g Tris base, 144 g glycine, and 10 g SDS in 1000 mL distilled water. Keep the buffer in cold for 2–3 h. Preferably add SDS before use. Dilute it to 10 times (1×).
3. Staining dye: Dissolve 1 g of Coomassie brilliant blue dye in 1 L of methanol: Water: glacial acetic acid solution (50:40:10). Filter it by Whatman paper 1 and store it at room temperature.
4. Destaining solution: Mixture of methanol: water: glacial acetic acid (50:40:10).

Protocols

1. **Extraction of proteins in RIPA buffer**: Described earlier.
2. **Protein quantitation**: Described earlier.
3. **Sample preparation**:

- After extraction, proteins are quantitated using an appropriate method.
- Take the required volume of the sample in an Eppendorf tube. Suppose 100 µg proteins to be taken for separation and the concentration of the sample is 5 µg/µL then take 20 µL from the sample.
- Now add the 6 × protein loading dye in a manner so that the final concentration of the dye becomes 1 ×.

  Calculation: N1V1 = N2V2

  6XV1 = 1 × (volume of the sample +V1)

  6XV1 = volume of the sample + V1

  6V1-V1 = volume of the sample

  5V1 = volume of the sample

  V1 = volume of the sample/5

  In the case of 20 µL of sample 4 µL of loading, the dye will be added.
- Mix well and incubate at 95°C for 15 min (for denaturation of proteins).
- Immediately keep the tube in the ice (to stop renaturation of proteins).

4. **Preparation of gel**:

   A. Resolving gel (10%):

Reagents	Stock	Working	Volume for 10 mL
30% acrylamide/bis solution	30%	10%	3.4 mL
10% SDS	10%	0.1%	0.1 mL
1.5 M Tris-Cl (pH 8.8)	1.5 M	375 mM	2.5 mL
10% APS	10%	0.1%	0.1 mL
TEMED			5 µL
Water			3.9 mL

The volume of 30% acrylamide/bis solution will be changed if the percentage of the gel changed as 12% or 15% depending on the size of the proteins.

B. Stacking gel (5%):

Reagents	Stock	Working	Volume for 5 mL
30% acrylamide/bis solution	30%	5%	0.833 mL
10% SDS	10%	0.1%	0.05 mL
0.5 M Tris-Cl (pH 6.8)	0.5 M	125 mM	1.25 mL
10% APS	10%	0.1%	0.05 mL
TEMED			5–6 µL
Water			2.9 mL

C. The casting of gel:
   1. The first set the apparatus and seal the glass plate (set the glass plate for gel casting).
   2. Pour the resolving or separation gel in the casting plate and leave some space above for stacking gel and comb.
   3. After pouring layer the above surface with water or isopropanol (to remove the bubbles and homogenous polymerization).
   4. Once the resolving gel gets polymerize, remove the isopropanol, and prepare the stacking gel solution.
   5. Pour the stacking gel solution on the separation gel and insert the comb immediately.
   6. After some time, remove the comb and clean the well for sample loading.

5. **Sample loading and gel running**:
   1. Clamp the plate with gel into the electrophoresis apparatus and fill the chamber with running buffer.
   2. Clean the well with diluted dye and load the prepared sample in the well as per experiment.
   3. Load the prestained protein marker if necessary.
   4. Connect the power supply and run at the lower voltage while moving through the stacking gel. Once all the proteins stack at the border of stacking/resolving gel, increase the voltage.

6. **Staining with CBB dye**:
   1. After the complete run, switch off the power supply and remove the gel carefully to a tray.
   2. Keep the gel in the CBB dye on the shaker for homogenous staining.
   3. After staining remove the dye and wash the gel 2–3 times with water.
   4. Destain the gel in a destaining solution.
   5. Visualize the bands.

Safety considerations and standards

1. All the tubes and glassware should be clean, free from reagent, and autoclaved.
2. Handle the acrylamide carefully as it is neurotoxic.
3. Prepare a fresh APS solution.
4. APS and TEMED should be added in the mix just before casting the gel.
5. Remove all the bubbles while casting the gel.
6. β-Mercaptoethanol is a harmful chemical. Avoid contact with skin and eyes.

7. SDS is neurotoxic. Dispose of all the waste as standard laboratory guidelines.
8. Avoid overloading of the sample in the well.
9. Wash all the apparatus after use.

Analysis and statistics

The proteins in the mixture are resolved on the gel based on their size and length, which can be visualized after staining with CBB.

Pros and cons

Pros	Cons
Used for protein separation purification and identification	Use of toxic chemicals as acrylamide, mercaptoethanol, etc.
Highly sensitive. Very less sample is required for separation and identification	Costly apparatus. Gel preparation is difficult
It can be used to determine the purity of the sample and the molecular weight of the purified protein	Proteins get denatured hence not suitable for studying protein-protein interaction

Troubleshooting and optimization

Problem	Solution
The sample is slimy and thick due to contamination of carbohydrates and nucleic acids	Lyse the cells properly, add endonuclease to digest DNA, precipitated protein in acetone/TCA
The sample does not settle in well	Increase the amount of glycerol in the loading buffer
Uneven polymerization of the gel	Pour the gel immediately after adding APS and TEMED
	Layer the upper surface of the resolving gel with water or isopropanol
Protein smear after staining	Due to proteolysis. Add protease and phosphatase inhibitors
Uneven staining	Stain on shaker
Poor resolution of the bands	Decrease the protein concentration
	Change the gel concentration
	Prepare fresh running buffers
Fuzzy bands	Remove any bubbles in the gel
	Load the sample properly and clean the well before loading
	Avoid the heating of the gel

Summary

1. The SDS-PAGE is a method of protein separation in the denaturing condition.
2. Polyacrylamide is used to prepare the gel or support medium. Bis-acrylamide cross-links the acrylamide polymer.
3. SDS binds the denatured protein (linear protein) and provides the overall negative charge thus, in SDS-PAGE the separation of the proteins is independent of the charge and structure. It is solely depending on the length or size of the protein.
4. In the stacking gel, the protein bands move with a slower rate and stack at the junction of stacking and resolving gel.
5. After separation, the bands can be visualized by staining with CBB dye.

Separation of Amino Acids by Paper Chromatography

Rationale

Paper chromatography is a technique used for the separation of compounds based on the differential solubility in the stationary phase and mobile phase (solvent). In the amino acid separation by paper chromatography, the solvent travel across the paper, it carries the amino acids with it. The amino acids are separated on the paper based on the differential solubility. R_f value (retention factor): Retention factor is the rate at which amino acid in the mixture travels along with the solvent. It is used to identify particular amino acids in the mixture during the chromatographic separation. Thus, after separation, the R_f value is calculated for each amino acid and identified by comparing it with the standard R_f value chart.

$$R_f = \frac{\text{Distance traveled by the substance (amino acid)}}{\text{Distance traveled by the solvent}}$$

Materials, equipment, and reagents

A. **Reagents**: Amino acids, isopropyl alcohol, ninhydrin, ethanol, butanol, glacial acetic acid, water.
B. **Glassware**: Petri dish, beaker, measuring cylinder, glass rod, toothpicks (capillary tube), Whatman paper, chromatographic tank.

Protocols

1. Take Whatman paper and cut it according to the size of the chromatographic tank.
2. Draw a line 1.5 cm away from the bottom edge.
3. Mark a spot on the line in the center.

4. Apply the amino acid solutions one by one on the spot with the help of a capillary tube.
5. Allow the spots to dry.
6. Fill the chromatographic tank (as an alternate beaker can be used) with solvent. Cover the bottom of the tank 0.5 cm depth.
7. Coil the spotted Whatman paper around a glass rod and tighten the end by stapler/clip. Adjust the paper in the tank in such a way as the only tip of the paper dipped in the solvent. The pencil line should be above the solvent.
8. Leave the arrangement for 2–3 h.
9. Remove the Whatman paper and mark the solvent front from the spot.
10. Dry the paper in the air.
11. Spray the ninhydrin solution over the paper and dry the paper again.
12. Each of the migrated amino acids on the paper develops a color on the paper. Mark the spots with pencil and measure the distance traveled by each spot.

Precursor techniques

1. Amino acid solutions: Dissolve 1 mg of each amino acid in 10 mL of 10% isopropyl alcohol.
2. Chromatographic solvent: Mix 250 mL N-butyl alcohol and 250 mL water in separating funnel. Add 60 mL glacial acetic acid in it. Mix well and keep for 15 min. Discard the lower layer and use the upper layer as a solvent.
3. Ninhydrin solution: Dissolve 0.2 g ninhydrin in 100 mL ethanol (90 mL acetone and 10 mL water).

Safety considerations and standards

1. Wear gloves while handling ninhydrin.
2. Avoid cross-contamination by using different stick/toothpick for each amino acid.
3. No heat or flames near the experiments as butanol is flammable.
4. Keep the Whatman paper in a vertical position in the chromatography tank.
5. Only the tip of the paper should be dipped in the solvent.
6. Handle the Whatman paper by top edge only.

Analysis and statistics

The distance traveled by each amino acid: RI, RII, RIII
The distance traveled by solvent front: RS

$$R_f \text{ value} = \frac{\text{Distance traveled by amino acid}}{\text{Distance traveled by the solvent}}$$

R_f value for amino acid I = RI/RS
R_f value of amino acid II = RII/RS
R_f value of amino acid III = RIII/RS

Pros and cons

Pros	Cons
1. Simple, easy, cheap, and rapid	1. Difficulty in separation if a greater number of amino acids present in the sample
2. Require very less material and good resolving power	2. Less accurate than other methods as HPLC

Alternative methods/procedures

Thin-layer chromatography (HPLC)

Troubleshooting and optimization

Problem	Solution
1. Thick spots of each amino acid after ninhydrin spray	Apply less amount of the mixture at the start on the paper
2. Nonlinear migration of the compounds	Dip the paper accurately

Summary

1. Amino acids can be separated in mixture by paper chromatography.
2. The amino acids in the mixture separated on the stationary phase based on the differential solubilities in the solvent (butanol: glacial acetic acid: water).
3. Each amino acid is identified by calculating the R_f value. R_f values are the rate at which amino acid migrates. It can be calculated by dividing the distance traveled by an amino acid with a distance of the solvent front.

CHAPTER 5

Mineral

1. Estimation of calcium in water or milk by EDTA titrimetric method.
2. Estimation of calcium in serum by O-cresolpthalein complexon method (colorimetric).
3. Estimation of the inorganic phosphorus by colorimetric method.
4. Estimation of the iron by thiocyanate colorimetric method.
5. Estimation of the chloride by titration method (Mohr's argentometric method).

ESTIMATION OF CALCIUM IN WATER OR MILK BY EDTA TITRIMETRIC METHOD

Definition

The hardness of water is mainly contributed by calcium and magnesium ions. The hardness of water is defined as total parts per million (ppm) of $CaCO_3$ by weight. For example, the hardness of water 10 ppm means 10 g of $CaCO_3$ in one million grams of water or 0.01 g of $CaCO_3$ in 1-L water. Total calcium concentration in water or milk can be determined by titration with ethylene diamine tetraacetic acid (EDTA), a calcium chelator.

Rationale

This is based on complexometric titration in which an indicator Eriochrome black T is added to the sample. The indicator forms a stable wine-red or pink complex with calcium. When this complex is titrated against EDTA, a more stable complex of Ca^{2+}-EDTA is formed. When the entire calcium ion form complex with EDTA, the free indicator turns the solution blue, which is an end point.

$$Ca + Indicator \rightarrow Ca - In\,(red/pink)$$

$$Ca - In + EDTA \rightarrow Ca - EDTA + Indicator\,(Blue)$$

Since magnesium is also present in the hard water, which also forms a similar complex, it should be removed by precipitation using NaOH/KOH.

Materials, equipment, and reagents

A. **Reagents**: EDTA, indicator (Eriochrome black T or pattons and readers), sodium hydroxide, dilute hydrochloric acid, sample.
B. **Glassware**: Burette, pipette, conical flask, volumetric flask, beaker, measuring cylinders.

Protocols

1. Fill the burette with 0.01 M EDTA.
2. Take 10 mL of sample in a conical flask.
3. Add 2–3 mL of 1-M NaOH solution (for precipitation of magnesium as magnesium hydroxide). Allow the solution to stand for 5–10 min.
4. Add 0.1 g of indicator and mix well.
5. Titrate it against 0.01 M EDTA solution.
6. Note the end point where the solution turns blue from red/pink.

Calculation

1. First, calculate the average volume of EDTA used for repeats (e.g., the reading of titrant used is 15 mL).
2. Now calculate moles of EDTA required to complex with calcium in the solution.

   The first way of calculation:

   An amount of 0.01 mol in 1000 mL of EDTA solution means 15 mL of EDTA contains 0.0015 mol. EDTA and calcium react in a 1:1 ratio, which means 1 mol of EDTA reacts with 1 mol of calcium. In addition, 10 mL of the sample solution contains 0.0015 or 0.15 mol/L (0.15 M). The molecular weight of calcium carbonate is 100 thus the concentration of the calcium carbonate is 15 g/L (15,000 ppm).

   The second way of calculation:

   The volume of the EDTA used (V1): V1 mL (let say 15 mL is used)

   Molarity of EDTA (M1): 0.01 M

   The volume of the sample taken for titration (V2): 10 mL

   Molarity of calcium carbonate (M2):

   M1V1 = M2V2

   $15 \times 0.01 = M2 \times 10$

   M2 = 0.15 M

   The molecular weight of calcium carbonate—100

$$Molarity \times Molecular\ weight = g/L$$

$$0.15 \times 100 = 15\,g/L$$

Precursor techniques

1. EDTA solution (0.01 M): Dissolve 3.723 g of disodium salt of EDTA ($Na_2EDTA \cdot 2H_2O$) in 1 L of water. Dissolve overnight if needed.

Protocols in Biochemistry and Clinical Biochemistry. https://doi.org/10.1016/B978-0-12-822007-8.00004-0

2. Eriochrome black T: It is a mixture of a one-part indicator and 99-part sodium chloride.
3. Sodium hydroxide solution (1 M): Dissolve 4 g of NaOH in 100 mL water.
4. Sample preparation: The calcium is present in tap water and milk. Solid samples are dissolved in dilute HCl.

Safety considerations and standards
1. Glassware should be washed before and after use.
2. The end point should be noted carefully.
3. For the estimation of only calcium in water, magnesium should be precipitated out completely before adding an indicator.

Analysis and statistics
The given solution contains _____ mg (ppm) of $CaCO_3$.

Pros and cons

Pros	Cons
It can be used for solid as well as liquid samples	Some metals like copper, iron, zinc, etc. if present in high concentration interferes with the reading
An easy and convenient method for calcium/ hardness water estimation	
No expensive instrument is required	

Alternative methods/procedures
O-Cresolpthalein complexon method for serum calcium, Baron, and Bell's titration technique for calcium.

Summary
1. It is a very easy and commonly used method for estimation of calcium in the sample.
2. In this complexometric titration method, calcium (present as calcium carbonate) is estimated by titrating with EDTA using Eriochrome black T as an indicator. Calcium and indicator form a red-pink complex. When this complex is titrated using EDTA, Ca-EDTA complex will form. At the end point, all the calcium complexed with EDTA and indicator alone will be in blue color.

ESTIMATION OF CALCIUM IN SERUM BY O-CRESOLPTHALEIN COMPLEXON METHOD (COLORIMETRIC)
Definition
Calcium can be estimated in a sample by flame photometry, titrimetric, colorimetric, fluorescent, etc.

O-Cresolpthalein complexon method is the colorimetric determination of calcium.

Rationale
O-Cresolpthalein complexon (CPC) reagent is a colorless indicator at neutral or acidic pH. In alkaline conditions, this reagent forms a red-purple complex with calcium. The absorbance of the complex is measured at 570 nm. The intensity of the color complex is directly proportional to the amount of calcium in the sample.

$$\text{Calcium} + O-\text{Cresolpthalein complexon} \xrightarrow{\text{Alkaline pH}} \text{Calcium} - O - \text{Cresolpthalein complexon (Red} - \text{purple color complex)}$$

8-hydroxyquinoline sulfonate is added to prevent magnesium and iron interference.

Materials, equipment, and reagents
A. **Reagents**: O-Cresolpthalein complexon reagent, diethylamine, calcium standard, 8-hydroxyquinoline sulfonate, sample.
B. **Glassware**: Beaker, test tube, measuring cylinders, cuvette.
C. **Instrument**: Colorimeter/spectrophotometer.

Protocols
1. Label three test tubes as B (blank), S (standard), and T (test).
2. Add the reagent and sample as described in the table.

	Blank (B)	Standard (S)	Test (T)
Standard calcium solution	–	0.1 mL	–
Sample	–	–	0.1 mL
CPC reagent	1 mL	1 mL	1 mL
Diethylamine	1 mL	1 mL	1 mL
Distilled water	0.1 mL	–	–
Absorbance (575 nm)	A1	B1	C1

Calculation

$$\text{Concentration of calcium}_{\text{(mg/dl)}} =$$
$$\frac{OD_T(C1) - OD_B(A1) \times \text{Concentration of standard}}{OD_S(B1) - OD_B(A1)}$$

The normal level of calcium in serum is 8.5–11 mg/dL.

Precursor techniques
1. Stock O-cresolpthalein complexon solution: Dissolve 40 mg of CPC in 1 mL of HCl. Water can be added to the dissolving process. Add 2.5 mg of 8-hydroxyquinoline sulfonate, mix well, and make the volume 1 L.

2. Diethylamine solution: Mix 3 mL of diethylamine solution in 97 mL of water.
3. Calcium standard: Dissolve 10 mg calcium in 100 mL of distilled water (0.1 mg/mL).

Safety considerations and standards
1. Reagents are toxic hence handle them carefully.
2. Since sodium azide is added to the reagents for stability, a lot of water should be used for its disposal.
3. Reagents should be stored at 2–8°C.
4. Serum samples with visible precipitate should not be used for the test.

Analysis and statistics
The given sample contains _____ mg/dL of calcium.

Pros and cons

Pros	Cons
A very easy and fast method for calcium estimation	Calcium-binding substances cause inaccurate results
	A high level of bilirubin also interferes with the result

Alternative methods/procedures
Baron and Bell's titration technique for calcium.

Summary
1. This colorimetric method can be used for the estimation of calcium in the serum sample.
2. O-Cresolpthalein complexon (CPC) reagent forms a red-purple complex with calcium at alkaline pH, which shows absorption maxima at 570 nm. The colorimetric measurement of its absorbance is proportional to the amount of calcium in the serum sample.

ESTIMATION OF THE INORGANIC PHOSPHORUS BY A COLORIMETRIC METHOD (FISKE AND SUBBAROW)

Definition
Phosphorus is an abundant mineral in the body that is present in the blood in organic as well as inorganic form. About 87% phosphorus in the body is present in bone. Inorganic phosphorus exists in the form of phosphate (H_3PO_4). The normal value of the inorganic phosphorus is 2.5–4.5 mg/dL.

Rationale
For the colorimetric estimation of inorganic phosphorus in the serum, proteins are first precipitate out with the use of trichloroacetic acid. The sample now treats with molybdic acid (ammonium molybdate and

H_2SO_4), which reacts with inorganic phosphorus to form phosphomolybdic acid. Phosphomolybdic acid is further reduced by reducing agent 1-amino-2-naphthol-4-sulfonic acid (ANSA) to molybdenum blue. The absorbance of the blue-colored solution is measured at 620 nm. The intensity of the blue-colored product is directly proportional to the amount of phosphorus present in the sample.

Reaction:

Ammonium molybdate + H_2SO_4 + Inorganic phosphorus
→ Phosphomolybdic acid

Phosphomolybdic acid + ANSA
→ Blue colored compound (OD at 620 nm)

Materials, equipment, and reagents
A. **Reagents**: Ammonium molybdate, sulfuric acid, 1-amino-2-naphthol-4-sulfonic acid (ANSA), sodium bisulfite, sodium sulfite, potassium dihydrogen phosphate (KH_2PO_4), a blood sample.
B. **Glassware**: Beaker, test tube, measuring cylinders, cuvette, wash bottle, volumetric flask, pipette.
C. **Instrument**: Colorimeter/spectrophotometer.

Protocols
1. Preparation of standard phosphorus solution (S): Prepare a series of standard dilutions of phosphorus by taking the different volumes of its working standard and make the volume 8.6 mL by adding distilled water.

Phosphorus Working Solution (11.39 µg/mL) (mL)	Distilled Water (mL)	Final Phosphorus Amount (µg)
1	7.6	11.39
2	6.6	22.78
3	5.6	34.17
4	4.6	45.56
5	3.6	56.95
6	2.6	68.34
7	1.6	79.73
8	0.6	91.12

2. For blank (B): take 8.6 mL of distilled water.
3. Take 1 mL of the test sample in the test tube (T) and add 7.6 mL of distilled water.
4. Add 1 mL of molybdate solution and 0.4 mL of ANSA solution in each of the standard test tubes, blank, and test samples.
5. Keep the test tubes at room temperature for 10 min for color development.
6. Measure the absorbance at 620 nm.
7. Plot the standard curve between absorbance and concentration of the standards.

8. Calculate the concentration of the unknown sample (test) from the standard plot.

Precursor techniques

1. Ammonium molybdate solution (2.5% in 3-N sulfuric acid): Dissolve 12.5 g of ammonium molybdate in 100 mL of water by heating. Cool it and add 160 mL of 10-N sulfuric acid. Make the final volume of 500 mL by distilled water.
2. 10-N sulfuric acid: Slowly add 100 mL of H_2SO_4 (36 N) in 260 mL of water in 500 mL beaker.
3. Sodium bisulfite solution (15%): Dissolve 30 g of sodium bisulfite in 200 of water.
4. Sodium sulfite (20%): Dissolve 10 g of sodium sulfite in 50 mL of distilled water. Filter if necessary.
5. 1-Amino-2-naphthol-4-sulfonic acid (ANSA) solution: Dissolve 0.5 g of ANSA in 195 mL of sodium bisulfite (15%) with heating (till 60°C). Cool it and add 5 mL of sodium sulfite solution (20%). Mix well.
6. Standard phosphorus solution: Dissolve 50 mg potassium dihydrogen phosphate (KH_2PO_4) in 1000 mL of distilled water.

   The molecular weight of KH_2PO_4—136
   The molecular weight of phosphorus—31

   Thus, 136 mg of KH_2PO_4 contains 31 mg of phosphorus. An amount of 10 mg KH_2PO_4 was dissolved in 100 mL of water. Also, 1000 mL solution has 11.39 ($31 \times 50/136$) mg phosphorus or 0.01139 mg/mL or 11.39 μg/mL.

Calculations

1. First, draw a standard curve between the standard concentration and corresponding absorbance.
2. Calculate the amount of iron in the unknown sample from the standard curve.

Safety considerations and standards

1. Dilution of H_2SO_4 is exothermic reaction hence acid should be added in water with stirring. It should be done very carefully.
2. Always measure absorbance from lower to a higher concentration.
3. Cuvette should be wiped from outside with tissue paper to remove traces of solution or dust before placing it in a cuvette holder.

Analysis and statistics

The given sample contains _____ mg/dL of inorganic phosphorus.

Alternative methods/procedures

Enzymatic method.

Summary

In the colorimetric estimation of inorganic phosphorus, a sample is treated with ammonium molybdate solution in the acidic condition. The resultant phosphomolybdic acid is further treated with ANSA and a molybdenum blue product is formed. The absorbance of the product measured at 620 nm. The intensity of the product is directly proportional to the inorganic phosphorus amount.

ESTIMATION OF THE IRON BY THIOCYANATE COLORIMETRIC METHOD

Definition

Iron is required by the body to perform various physiological functions. Iron forms complex with heme protein from hemoglobin. This complex is present in red blood cells, carries oxygen in the blood. The deficiency of hemoglobin leads to an anemic condition. Anemia also caused by a deficiency of iron in the diet. Myoglobin and cytochromes are the other complex where iron acts as a cofactor. Iron is required in little amount though it plays an important role in the functioning of the body.

Rationale

In the colorimetric estimation, iron first reduced to ferric ion (Fe^{3+}) by some oxidizing agents like potassium persulfate and hydrogen peroxide. The ferric ion forms a visible blood-red complex with thiocyanate (ferric thiocyanate). The intensity of the color depends on the amount of iron present. The absorbance of the solution can be measured at 490 nm.

$$Fe^{3+} + SCN^- \rightarrow (FeSCN)^{2+}$$

The concentration of the iron now calculated using the absorbance with a different standard solution of the iron (Fe^{3+}).

Materials, equipment, and reagents

A. **Reagents**: Ferrous ammonium sulfate, potassium thiocyanate, potassium permanganate ($KMnO_4$), concentrated sulfuric acid, sample.
B. **Glassware**: Glass test tube, test tube stand, conical flask, pipette, beakers, measuring cylinder, funnel, cuvette, volumetric flask.
C. **Instruments**: Spectrophotometer, pan balance.

Protocols

1. Preparation of standard solutions in different test tubes:

Fe³⁺ Standard Solution	Distilled Water (mL)	Amount of Ferric Ion (mg)
0.2 mL (S1)	1.8	0.02
0.4 mL (S2)	1.6	0.04
0.6 mL (S3)	1.4	0.06
0.8 mL (S4)	1.2	0.08
1 mL (S5)	1	0.1

2. Colorimetric estimation of absorbance:
Take other test tubes and label as blank (B), test (T).

Reagents	Blank	Standard (S1-S5)	Test
Standard solution	–	As mentioned in the above table	–
Test sample	–	–	1 mL
Distilled water	2 mL	–	1 mL
Conc. sulfuric acid	0.5 mL	0.5 mL	0.5 mL
Pot. persulfate	1 mL	1 mL	1 mL
Ammonium thiocyanate	2 mL	2 mL	2 mL
	Keep it for 5 min for color development		
OD at 490 nm	OD_B	OD_{S1-S5}	OD_T

Precursor techniques

1. Preparation of Fe^{3+} standard solution: Dissolve 702-g ferrous ammonium sulfate in 1000 mL of distilled water. Add 5 mL of conc. sulfuric acid. Warm it and add 5 mL of potassium permanganate dropwise. Potassium permanganate oxidizes ferrous to ferric ion. Now make the final volume again 1000 mL.

    This solution contains 0.1 mg iron (Fe^{3+}) per mL.

    The molecular weight of ferrous ammonium sulfate is 392.

    The molecular weight of iron is 56.

    An amount of 392-mg ferrous ammonium sulfate contains 56 mg of ferrous (iron).

    Thus, 702-mg ferrous ammonium sulfate contains 100 mg of iron.

    Since it is dissolved in 1000 mL of distilled water, 100 mg iron in 1000 mL water—0.1 mg/mL.
2. Preparation of ammonium thiocyanate solution (1 M): Dissolve 7.6-g ammonium thiocyanate in 100 mL of distilled water.
3. Preparation of potassium permanganate solution (0.15 M): Dissolve 2.4-g potassium permanganate in 100 mL of distilled water.
4. Saturated potassium persulfate: Dissolve 7-g potassium persulfate in 100 mL of distilled water.

Calculations

1. First, draw a standard curve between the standard concentration and corresponding absorbance.
2. Calculate the amount of iron in the unknown sample from the standard curve.

Safety considerations and standards
1. Glassware should be washed before and after use.
2. Handle the chemicals carefully.

Analysis and statistics
The amount of iron in the given sample is _____ mg/dL.

Alternative methods/procedures
Helimeyer's method, the Peterson method, etc.

Summary
In this colorimetric estimation of iron, the sample is first treated with oxidizing agents like persulfate to oxidize ferrous to ferric ion. The ferric ion (Fe^{3+}) then reacts with thiocyanate to form a red complex of ferric thiocyanate, which absorbs at 490 nm. Its intensity is the direct measurement of the iron concentration.

ESTIMATION OF THE CHLORIDE BY TITRATION METHOD (MOHR'S ARGENTOMETRIC METHOD)
Definition
Chloride is an essential electrolyte in the human body, which involved in metabolism and acid-base balance along with other electrolytes as sodium, potassium, etc. The maximum allowable concentration of chloride in drinking water is 250 mg/L.

The normal value of chloride in blood is 97–107 mEq/L. Hyperchloremia is associated with dehydration, high blood sodium, kidney disease, etc.

Rationale
The sample (with chloride) is titrated against silver nitrate. Potassium chromate is added as an indicator. Silver nitrate reacts with chloride ions to form insoluble white silver chloride. Once all the chloride consumes, the excess silver combines with potassium chromate to form an orange-red-colored compound, which is the end point of the titration.

$$AgNO_3 + K_2CrO_4 + Cl^- \rightarrow AgCl_2 + NO^{3-} + K_2CrO_4$$

$$2AgNO_3 + K_2CrO_4 \rightarrow Ag_2CrO_4(Orange-red) + 2KNO_3$$

Materials, equipment, and reagents
A. **Reagents**: Silver nitrate, potassium chromate indicator.
B. **Glassware**: Burette, stand, volumetric flask, pipette, conical flask, measuring cylinders.

Protocols

1. Dilute the sample in distilled water.
2. Take 25 mL of the sample in a conical flask.
3. Add 1 mL of potassium chromate indicator.
4. Fill the burette with 0.1-M silver nitrate titrant.
5. Titrate the sample against it.
6. The end point occurs with the appearance of orange-red color. Note down the titrant used. Repeat the titration till two concordant readings.

Precursor techniques

1. Silver nitrate (0.1 M): Dissolve 9 g of $AgNO_3$ in 500 mL of distilled water. Store the solution in a brown bottle.
2. Potassium chromate indicator (0.25 M): Dissolve 0.5 g of potassium chromate in 10 mL of distilled water.

Calculations

1. Since silver nitrate and chloride are reacting in 1:1 mole ratio

$$Ag^{2+} + Cl^- \rightarrow AgCl$$

First, calculate the molarity of the chloride ions
M1V1 = M2V2
M1 = M2V2/V1
M1 = Molarity of chloride ion (sample)
V1 = Value of the sample
M1 = Molarity of the titrant (silver nitrate)—0.1 M
V2 = Volume of the titrant used

2. Now calculate the concentration of chloride ions
Molecular weight of chloride—35.5

$$\text{Chloride ion concentration (mg/L)} = \frac{M2V2 \times 35.5 \times 1000}{V1}$$

Safety considerations and standards

1. The pH of the sample should be between 6.5 and 10. For acidic solutions, gravimetric or Volhard's methods to be used.
2. Do not add indicator in excess.

Analysis and statistics

Color changes to orange-red are the indication of end point in the titration. The concentration of the chloride can be calculated as the above formula.

Pros and cons

Pros	Cons
Very easy titration method Simple and accurate method	Interference by bromide ion in the sample
Very less time required	Excess titrant (silver nitrate) is needed
The potassium chromate indicator is an easily accessible and cheap reagent	Sodium carbonate also interferes by reacting with silver nitrate to form silver carbonate precipitate. This can be eliminated by adding a small amount of sulfuric acid

Alternative methods/procedures

Mercuric thiocyanate method (colorimetric), Volhard's method (titration), mercuric nitrate method (titration).

Summary

1. Mohr's method is a titrimetric method of chloride estimation using silver nitrate as a titrant.
2. Silver nitrate reacts with the chloride to form a white precipitate of silver chloride and at end point excess silver nitrate form an orange-red complex (silver chromate) with indicator potassium chromate.
3. The chloride ion concentration calculated as 1 mol of silver nitrate combine with 1 mol of chloride.

CHAPTER 6

Vitamin

1. Estimation of vitamin A by trichloroacetic acid (modified Carr-Price method).
2. Estimation of the tocopherol (vitamin E) by colorimeter (Baker and Frank method).
3. Estimation of the thiamine (vitamin B1) by thiochrome technique.
4. Estimation of the ascorbic acid (vitamin C) by 2,6-dichlorophenol indophenol (DCPIP) method (titrimetric).
5. Estimation of the ascorbic acid (vitamin C) by 2,4 dinitrophenylhydrazine method (colorimetric).

ESTIMATION OF VITAMIN A BY TRICHLOROACETIC ACID (MODIFIED CARR-PRICE METHOD)

Definition

Retinol (vitamin A alcohol), retinaldehyde (vitamin A aldehyde), retinoic acid (vitamin A acid), and their analogs are grouped under retinoids. Carotenes are the precursor for the synthesis of vitamin A. Vitamin A is required for the maintenance of and normal functioning of the eye, bone development, and epithelial lining of the reproductive and respiratory tube. Vitamin A is lipid soluble and stored in liver tissues and its deficiency causes night blindness and xerophthalmia.

Rationale

Trichloroacetic acid (TCA) in chloroform reacts with diluted vitamin A to form a transient blue-colored compound. The intensity of the color produced is directly proportional to the amount of vitamin A. The absorbance of the compound can be measured at 620 nm. The normal serum vitamin A varies between 15 and 60 μg per 100 mL.

Materials, equipment, and reagents

A. **Reagents**: Trichloroacetic acid (TCA), standard vitamin A, chloroform, sample.
B. **Glassware**: Beaker, measuring cylinder, test tube.
C. **Instrument**: Spectrophotometer/colorimeter.

Protocols

1. From the standard vitamin A solution (10 μg/mL), make different concentrations of standards (1–7 μg) in 1 mL of chloroform.

Vitamin A Solution (μL) (10 μg/mL)	Chloroform	Amount of Vitamin A Per mL (μg)
0	1 mL	0 (blank)
100	900 μL	1
200	800 μL	2
300	700 μL	3
400	600 μL	4
500	500 μL	5
600	400 μL	6
700	300 μL	7

2. Add 2 mL of TCA reagent to each test tube rapidly, mix well and take the absorbance or optical density at 620 nm using a colorimeter or spectrophotometer.
3. Plot a standard curve between absorbance and vitamin A standard solution.
4. Take 1 mL of the sample and add 2 mL of TCA reagent, mix well, and take the absorbance or optical density at 620 nm using a colorimeter or spectrophotometer setting the absorbance zero with blank sample (TCA).

Calculation

From the standard curve, the concentration of vitamin A in an unknown sample can be calculated.

Precursor techniques

1. Vitamin A solution: Dissolve 1-mg vitamin A acetate in 100 mL of chloroform (10 μg/mL).
2. TCA solution: Dissolve 50 g of TCA crystals in alcohol-free 25 mL of chloroform. This yields 5 mL of reagent. Keep this solution away from light and use it within 5 h of preparation.
3. Sample preparation: 15 mL of serum is mixed with 15 mL of ethanol. Add 30 mL of light petroleum and mix using a mechanical shaker for 1–2 min to obtain a supernatant solution. 25 mL of the aliquot is evaporated under vacuum, and the residue is dissolved in 1.25 mL chloroform. Use 1 mL of the solution (equal to 5 mL of serum sample) for further use.

Protocols in Biochemistry and Clinical Biochemistry. https://doi.org/10.1016/B978-0-12-822007-8.00003-9

Safety considerations and standards

1. The sample and reagents must be kept away from light.
2. TCA solution should be freshly prepared.

Analysis and statistics

The given serum sample contains _____ μg vitamin A in 1 mL.

Pros and cons

Pros	Cons
TCA reagent is easily available and easy to handle	The blue color is not stable
TCA is easy to dissolve in chloroform and is less toxic than TFA and SbCl$_3$ (antimony trichloride)	This method does not differentiate between cis and trans-form, thus it is not specific for all trans-form

Alternative methods/procedures

Determination of vitamin A by conversion to anhydro-vitamin A (Budowski and Bondi, 1957). Estimation of vitamin A using glycerol 1,3-dichlorohydrin (Sobel and Werbin, 1945).

Summary

This is a colorimetric estimation of vitamin A. Vitamin A forms a blue-colored compound when it reacts with TCA in chloroform solution. The intensity of the color complex is directly proportional to the amount of vitamin A in the sample. The absorbance of the final solution can be measured at 620 nm.

ESTIMATION OF THE TOCOPHEROL (VITAMIN E) BY COLORIMETER (BAKER AND FRANK METHOD) (OXIDIMETRIC COLOR REACTION)

Definition

Vitamin E is a fat-soluble vitamin. It is an important antioxidant in the body and regulates various diseases like hypertension, cardiac disease, male fertility, etc. The normal range of vitamin E in the blood is 5.5–17 mg/L.

Rationale

α-Tocopherol reduces the ferric ion into ferrous ion (in Emmerie-Emmerie Engel reaction), which then reacts with α,α′-dipyridyl to form a red-colored complex. The absorbance of which can be measured at 520 nm.

To remove the carotene interference, tocopherol and carotenes are extracted in xylene and the absorbance for carotenes is measured at 460 nm. After adding ferric chloride read the absorbance at 520 nm. Subtract the reading at 460 from the reading at 520 nm.

Materials, equipment, and reagents

A. **Reagents**: Absolute ethanol (aldehyde free), xylene, α,α′-dipyridyl, ferric chloride, α-tocopherol tablets, n-propanol.
B. **Glassware**: Beaker, measuring cylinder, test tube, cuvette.
C. **Instrument**: Spectrophotometer/colorimeter.

Protocols

Take the three test tubes and label them as blank (B), standard (S), and test (T).

Add the following reagents as shown in the table:

Reagents	Blank	Standard	Test
Standard tocopherol	–	1.5 mL	–
Serum sample	–	–	1.5 mL
Water	1.5 mL	1.5 mL	–
Ethanol	1.5 mL	–	1 mL
Xylene	1.5 mL	1.5 mL	1.5 mL
Centrifugation at 1500 rpm for 10 min			
Transfer the upper xylene layer in the separate test tube			
Supernatant xylene layer	1 mL	1 mL	1 mL
α,α′-Dipyridyl	1 mL	1 mL	1 mL
The absorbance of standard and test solution at 460 nm against the blank (set zero using blank)			
Ferric chloride	0.33 mL	0.33 mL	0.33 mL
The absorbance of standard and test solution at 520 nm against the blank (set zero using blank)			

Calculation

Concentration of vitamin E (mg/L)

$$= \frac{\text{Abs of test (520nm)} - (\text{Abs of the test (460nm)} \times 0.29)}{\text{Abs of standard (520nm)}}$$
$$\times \text{Concentration of the standard (10)}$$

Precursor techniques

1. α,α′-Dipyridyl: Dissolve 120 mg α,α′-dipyridyl in 100 mL n-propanol.
2. Ferric chloride solution: Dissolve 120 mg FeCl$_3$·6H$_2$O in 100 mL ethanol.
3. Standard vitamin E (tocopherol) solution (10 mg/L): Dissolve 100 mg of vitamin E tablet in 100 mL ethanol. Take 10 mL of it and make the volume 1000 mL by ethanol.

Safety considerations and standards

1. While transferring the xylene layer, do not take any ethanol and protein part.
2. Weighing should be done accurately.

Analysis and statistics

The given serum sample contains _____ mg per mL vitamin E.

Alternative methods/procedures

Martinek's (1964) method.

Summary

In the colorimetric estimation method of tocopherol, the ferric ion is reduced to ferrous ion by tocopherol. The amount of ferrous ion is directly proportional to the tocopherol concentration. The reduced ferrous ion then from a red complex with α,α'-dipyridyl. Thus, the intensity of this complex is in direct measurement of the tocopherol that reduces Fe^{3+} to Fe^{2+}.

ESTIMATION OF THE THIAMINE (VITAMIN B1) BY THIOCHROME TECHNIQUE (FLUORIMETRIC)

Definition

Vitamin B1 is a water-soluble vitamin. Thiamine pyrophosphate (TPP) is the active form of vitamin B1. Vitamin B1 acts as a cofactor of many enzymes. Thiamine is involved in nerve conduction, muscular functioning, metabolism, etc. The deficiency of thiamine causes beriberi disease, appetite loss, and neuritis.

Rationale

Thiamine is oxidized to a fluorescent compound, thiochrome by alkaline potassium ferricyanide. The fluorescence of the thiochrome is measured by fluorimeter. The normal serum vitamin B1 varies between 2.5 and 7.5 µg per 100 mL.

Materials, equipment, and reagents

A. **Reagents**: Potassium chloride, sodium hydroxide, potassium ferricyanide, standard thiamine solution, conc. HCl, isobutyl alcohol, sodium sulfate, sample.
B. **Glassware**: Beaker, measuring cylinder, test tube.
C. **Instrument**: Fluorometer.

Protocols

1. Take 5 mL of thiamine standards (0.1–1 µg/mL) in different test tubes.
2. Add 3 mL of NaOH (15%) in each of the test tubes.
3. Add 100 µL potassium ferricyanide and mix well for the oxidation process.

4. Add 15 mL of isobutanol. Mix well for 1 min allowing the tubes to stand till two layers separate.
5. Remove the aqueous layer carefully.
6. The isobutanol layer contains thiochrome. To this add 2–3 g of sodium sulfate and mix well. It will absorb the water and the solution becomes clear.
7. Measure the fluorescence of the clear solution using a fluorometer.
8. Prepare sample blank by using 5 mL of 25% KCl and repeat all the steps except the addition of potassium ferricyanide. Set zero of fluorometer using this blank solution.
9. Take 5 mL of the test sample and repeat all the steps and measure fluorescence.

Calculation

From the fluorescence of the concentration of the standard solution of the vitamin B1 in an unknown sample can be calculated.

Precursor techniques

1. Sodium hydroxide (15%): Dissolve 15 g NaOH in 100 water.
2. Potassium ferricyanide (1%): Dissolve 1 g of potassium ferricyanide in 100 mL of water.
3. Potassium chloride (25%): Dissolve 250 g of potassium chloride in 1 L of 0.1 N HCl.
4. Standard thiamine solution:
   (A) Stock: Dissolve 10 µg of thiamine in 10 mL of 25% KCl (1 µg/mL).
   (B) Working: Serial dilution of stock solution in 25% KCl (0.1–1 µg/mL).

Safety considerations and standards

1. Handle the HCl carefully. HCl is highly toxic. Dilute it by adding acid in water.
2. Solutions should be freshly prepared.

Analysis and statistics

The given serum sample contains vitamin B1_____ µg/100 mL.

Alternative methods/procedures

Determination of vitamin B1 by the spectrophotometric method.

Summary

This is the fluorometric method. Vitamin B1 is oxidized by potassium ferricyanide into a fluorophore thiochrome whose fluorescence is the direct measurement of the thiamine or vitamin B1 in the sample.

ESTIMATION OF VITAMIN C (ASCORBIC ACID) BY INDOPHENOL METHOD IN A GIVEN SOLUTION (TITRIMETRIC METHOD)

Definition

Vitamin C is a water-soluble vitamin that plays a vital role in the repair and development of tissues. It is involved in collagen synthesis, wound healing, and in immunity. Its deficiency leads to a condition known as Scurvy characterized by bleeding gums, joint, and muscle pain. Citrus fruits and green vegetables are a rich source of vitamin C.

Rationale

2,6-Dichlorophenolindophenol (DCPIP) is a blue dye which converts to pink color in acidic condition. Ascorbic acid is a strong reducing agent and strong acid. In the oxidation-reduction reaction, when ascorbic acid is titrated against blue dye DCPIP, it oxidized to dehydroascorbic acid (DHA) reducing dye to colorless leucobase. Oxalic acid is used in titrating medium. In a titration, end point occurs when all the ascorbic acid is converted to DHA, and further addition of ascorbic acid converts blue dye into pink color in the acidic condition.

$$L - Ascorbic\ acid + DCPIP \rightarrow L - dehydroascorbic\ acid\ (DHA) + Leucobase$$

Materials, equipment, and reagents

A. **Reagents**: 2,6-Dichlorophenolindophenol (DCPIP) dye, sodium bicarbonate, ascorbic acid, oxalic acid.
B. **Glassware**: Burette, burette stand, conical flask, beaker, measuring cylinder.

Protocols

Standard titration:
1. Fill the burette with standard DCPIP dye.
2. Take 5 mL of standard ascorbic acid standard solution (1 mg/mL) in a 10 mL conical flask.
3. Titrate ascorbic acid against the dye till the end point is reached with the appearance of pink color. (The volume of dye used is V1.) Repeat the titration till two same V1 reading is obtained. The amount of dye used is equal to the amount of ascorbic acid.

Sample titration:
1. Take 10 mL of the given unknown solution in a 100 mL conical flask.
2. Add 5 mL of oxalic acid (4%) in it.
3. Titrate it against dye till light pink color appears.
4. Note the volume of the dye V2.
5. Calculate the amount of ascorbic acid.

Calculation

Strength of standard ascorbic acid: 1 mg/mL.
The volume of the standard ascorbic acid solution: 5 mL.
The volume of dye solution required: V1.
V1 volume of dye solution is reduced by 5 mL (1 mg/mL) or 5 mg ascorbic acid.

In the sample titration suppose V2 volume of the dye is used to reduce the ascorbic acid in the unknown sample then

Amount of ascorbic acid in unknown sample $= (5 \times V2)/V1$

If the value of the V1 is 10 mL and V2 is 16 mL then the amount of ascorbic acid is $5 \times 16/10 = 8$ mg. Since the volume of an unknown sample taken is 10 mL. Concentration is 8 mg/10 mL or 80 mg/dL.

Precursor techniques

1. Standard dye solution: Dissolve 42 mg of sodium bicarbonate and 52 mg of 2,6-dichlorophenolindophenol (DCPIP) dye in 50 mL of water. Heat the solution if the dye does not dissolve. Make the volume 200 mL and filter it. Store the dye at 4°C. In alkaline conditions, the dye is blue in color.
2. 4% Oxalic acid: Dissolve 4 g of 100 mL of water.
3. Standard ascorbic acid: Dissolve 100 mg ascorbic acid in 100 mL (1 mg/mL) of 4% oxalic acid.

Safety considerations and standards

1. The standard dye solution should be prepared fresh.
2. Oxidation of ascorbic acid should be avoided while preparing it.

Analysis and statistics

The given solution contains _____ mg/dL of ascorbic acid.

Pros and cons

Pros	Cons
No requirement for expensive instruments	Interference by other reducing agents
Easy method	

Alternative methods/procedures

Colorimetric estimation by 2,4-dinitrophenylhydrazine, bromine method, iodometric titration method.

Summary

1. This is the titrimetric method of vitamin C estimation.
2. This method is based on the principle that ascorbic acid (vitamin C) is a strong reducing agent and when

titrated with blue DCPIP dye, it reduces the dye and gets oxidized. Oxalic acid is added to the solution. When all the ascorbic acid is reduced, the extra dye in the acidic condition converts to the pink color which is the end point of the titration. The ascorbic acid is estimated in the unknown sample by standardizing with a standard ascorbic acid solution.

ESTIMATION OF VITAMIN C (ASCORBIC ACID) BY 2,4-DINITROPHENYLHYDRAZINE METHOD IN A GIVEN SOLUTION (COLORIMETRIC METHOD)

Rationale

Ascorbic acid becomes oxidized by the copper sulfate solution. When the oxidized product dehydroascorbic acid is treated with 2,4-dinitrophenyl hydrazine, its ketone group reacts and forms dinitrophenylhydrazine (osazone). The osazone gives a red-colored complex in strongly acidic conditions (sulfuric acid). The intensity of the red color is directly proportional to the amount of the ascorbic acid which can be measured at 540 nm.

Ascorbic acid + Copper sulfate → Dehydroascorbic acid

Dehydroascorbic acid

+ 2,4 dinitrophenyl hydrazine $\xrightarrow{\text{Sulphuric acid}}$ Osazone

(pale yellow to red, 540 nm)

Materials, equipment, and reagents

A. **Reagents**: 2,4-dinitrophenylhydrazine reagent, thiourea, sulfuric acid, copper sulfate, trichloroacetic acid (TCA), acetic acid, metaphosphoric acid, ascorbic acid.

B. **Glassware**: Test tube, volumetric flask, beaker, measuring cylinder, stand, cuvette.

C. **Instrument**: Spectrophotometer/colorimeter.

Protocols

1. Preparation of standard dilutions: Take five test tubes and label them as S1, S2, S3, S4, and S5.

Standard Solution (mL)	5% TCA (mL)	Amount of Ascorbic Acid
0.5	2.5	0.5 g
1	2	1 mg
1.5	1.5	1.5 mg
2	1	2 mg
2.5	0.5	2.5 mg

2. Take two test tubes and label as blank (B) and test (T).

Reagents	Blank	Standards	Test
Standard ascorbic acid	–	As prepared above	–
Sample	–	–	1 mL
TCA	3 mL	–	2 mL
DTC reagent	1 mL	1 mL	1 mL
	Incubate at 60°C for 1 h		
	Cool on ice for 15 min		
9 N sulfuric acid	5 mL	5 mL	5 mL
	Vigorous shaking and keep at room		
	temperature for 15 min		
	OD at 540 nm		

Calculation

1. First plot a standard curve between the concentration of standard and corresponding absorbance.
2. Calculate the amount of ascorbic acid in the test sample from the standard curve. Use a blank sample to set zero absorbance in the colorimeter or substrate the blank absorbance from the standard and test absorbance.
3. Multiply the amount by 100 to get the amount in 100 mL sample.

Precursor techniques

1. 2,4-Dinitrophenylhydrazine reagent: Dissolve 2 g of 2,4 dinitrophenylhydrazine reagent in 100 mL 9 N sulfuric acid. Keep it overnight.
2. 0.6% Copper sulfate: Dissolve 0.6-g copper sulfate in 100 mL distilled water.
3. 5% Trichloroacetic acid.
4. 5% Metaphosphoric acid in 10% acetic acid.
5. Dinitrophenylhydrazine-thiourea-copper sulfate (DTC) reagent (color reagent): Mix 5 mL of thiourea, 5 mL of copper sulfate, and 100 mL DNPH reagent.
6. Standard ascorbic acid solution (1 mg/mL): Dissolve 100 mg ascorbic acid in 100 mL of 5% metaphosphoric acid in 10% acetic acid solution.

Safety considerations and standards

1. Reagents should be freshly prepared.

Analysis and statistics

The given solution contains _____ mg/dL of ascorbic acid.

Alternative methods/procedures

Titrimetric method by indophenol, bromine method, iodometric titration method.

Summary

1. This is the colorimetric method of vitamin C estimation.

2. In this method oxidized ascorbic acid (by copper sulfate) forms osazone with 2,4-dinitrophenylhydrazine which further converts to a red-colored complex in strong acidic condition. The absorbance can be measured at 540 nm.

CHAPTER 7

Enzymes

1. To study the effect of pH and temperature on the action of salivary amylase.
2. Determination of the activity of amylase enzyme (crude) extracted from sweet potato.
3. Determination of the specific activity of the amylase enzyme.
4. To determine V_{max} and K_m in of the amylase enzyme.

They are biological catalysts and are needed in very small quantities to carry out their function. They increase the rate of reaction by decreasing the activation energy. They do not affect the final equilibrium of the reaction. The specificity of the enzymes is very narrow, i.-e., they can catalyze a very small range of reactions and in most cases, a single reaction. In this chapter, we discuss some points regarding the activity of the enzymes. They are quite large molecules as compared to the substrates and bind to them by the means of the active site that is formed by folding of the protein in a three-dimensional manner.

MEASURING THE ENZYME ACTIVITY

Enzyme Assay
They are assayed by tracking the appearance of the product or disappearance of the substrate with the progress of the time.

Progress curve
A plot that represents the amount of product formed or the substrate changed with time is called the progress curve.

In the initial phase of the reaction, it is linear and falls with the progress of the reaction.

Enzyme activity is obtained from the initial phase of the curve, the linear part, the initial velocity of the reaction:

$$\vartheta_0 = \frac{a}{b}$$

Enzyme activity is expressed in terms of units (U), and one unit of enzyme is the amount of enzyme that can convert $1\,\mu M$ of the substrate to its products in 1 min, under defined conditions. Its SI unit is katal (kat) and is measured by the amount of enzyme required to convert 1 mol of substrate per second.

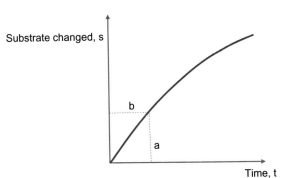

Progress curve—the initial activity of the enzyme is the slope of the linear part of the progress curve, $v=a/b$.

Enzyme kinetics
The enzyme-catalyzed reaction can be represented by the following equation:

$$E + S \underset{k_{-1}}{\overset{k_{+1}}{\rightleftharpoons}} ES \overset{k_2}{\rightarrow} E + P$$

where
$E=$ enzyme concentration
$S=$ substrate concentration
$P=$ product concentration
$k_{+1}, k_{-1},$ and $k_2 =$ rate constant

For a fixed concentration of enzymes, the initial rate, ϑ_0 is a hyperbolic manner with substrate concentration. This relationship can be mathematically expressed as the Michaelis-Menten equation.

$$\vartheta_0 = \frac{V_{max}[S]}{K_m + [S]} \tag{1}$$

where
$\vartheta_0=$ initial rate of the reaction
$V_{max}=$ maximum velocity of the reaction
$S=$ substrate concentration
$K_m=$ Michaelis-Menten constant.

$$\text{Michaelis} - \text{Meten constant}, K_m = \frac{k_2 + k_{-1}}{k_{+1}}$$

Generally, k_2 is very small and can be ignored, then K_m is equal to the dissociation constant.

Protocols in Biochemistry and Clinical Biochemistry. https://doi.org/10.1016/B978-0-12-822007-8.00002-7

Thus, the larger the K_m lower will be the affinity of the enzyme toward the substrate, and the smaller the K_m more the affinity of the enzyme toward the substrate and vice versa.

V_{max} is the maximum velocity of the reaction when all the sites of the enzymes are occupied. When the substrate concentration is low, all the sites on the enzymes will not be occupied and the rate of the reaction will be dependent on the concentration of the substrate and follows the first-order kinetics. However, when the substrate concentration is too high, all the active sites of the enzymes are occupied by the substrate and the rate of the reaction is independent of the substrate concentration and is the zero-order reaction.

When the velocity of the reaction reaches half of the max velocity, Eq. (1) can be rearranged as

$$\frac{V_{max}}{2} = \frac{V_{max}[S]}{K_m + [S]}$$

or

$$K_m = 2[S] - [S] \tag{2}$$

or

$$K_m = [S]$$

The use of the Michaelis-Menten equation is not popular to determine the kinetic constants as it is difficult to measure the initial rate at high substrate concentration and then extrapolating the hyperbolic curve. Therefore,

its linear transformation like the Lineweaver-Burk plot is commonly used. It is the reciprocal of the Michaelis-Menten equation.

$$\frac{1}{\vartheta_0} = \frac{1}{V_{max}} + \frac{K_m}{V_{max}} \frac{1}{[S]} \tag{3}$$

Plotting $1/\vartheta$ against $1/S$ gives a straight line with slope K_m/V.

When

$$\frac{1}{\vartheta_0} = 0, \quad \frac{1}{[S]} = \frac{1}{K_m}$$

and

$$\frac{1}{[S]} = 0, \quad \frac{1}{\vartheta_0} = \frac{1}{V_{max}}$$

Alternatively, multiplying the Lineweaver-Burk equation by S will give Hanes equation (S/ϑ_0) plotted against S:

$$\frac{[S]}{\vartheta_0} = \frac{[S]}{V_{max}} + \frac{K_m}{V_{max}}$$

Rearranging the Michaelis-Menten equation will give the following equation:

$$\vartheta_0 = V_{max} - \frac{K_m \vartheta_0}{[S]}$$

This is the Eadie-Hofstee equation, ϑ_0 verses $\vartheta_0/[S]$ can be plotted using this.

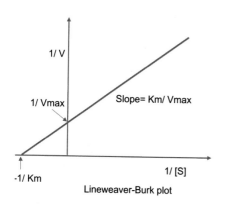

Lineweaver-Burk plot

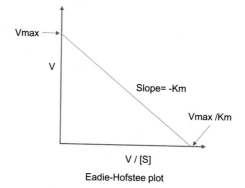

Eadie-Hofstee plot

Definition

To study the effect of pH and temperature on the action of salivary amylase.

Rationale

Enzymes are the biological catalysts and are known to speed up the rate of a reaction without affecting the substrate and products. Enzymes remain unaltered at the end of the reaction. There are various factors, such as pH, temperature enzyme concentration, substrate concentration, etc., that affect the activity of an enzyme. The optimal activity of an enzyme is generally between pH 5.0 and 9.0, while that in case of temperature is approximately 37°C. Various enzymes work best beyond this range, e.g., pepsin that works at pH 2. In this experiment, the saliva containing the amylase enzyme breaks down the starch. The optimum temperature for amylase is 37°C and pH 6.8. The chloride ion induces the allosteric activation of salivary amylase. The orange-brown iodine solution turns blue when starch is added to it, therefore, when the 2 drops of control are added to the iodine, the color of the solution changes depending on the availability of the starch.

Materials, equipment, and reagents

A. **Reagents**: 1% NaCl, 1% starch solution, 1% iodine solution, pH buffers obtained by dissolving pH tablets in water.
B. **Glassware**: Test tube, test tube holder, dropper, test tube stand, beaker.
C. **Instrument**: Water bath.

Protocol

For pH
1. Take three sets of test tubes with eight tubes in each set. Mark the test tube stands as pH 4, pH 6.8, and pH 9.
2. In each test, take 0.5 mL of iodine solution (orange-brown).
3. Take three separate test tubes and in one beaker take 15 mL of 1% starch solution, 3 mL of 1% NaCl solution, and divide it into the three test tubes equally.
4. Add 1 mL of pH 4 buffer and mark it as the control for stand A; add 1 mL of pH 6.8 buffer and mark it as the control for stand B; and add 1 mL of pH 9 buffer and mark it as the control for stand C.
5. Transfer all three test tubes into a water bath set at 37°C.
6. In a beaker, collect saliva. Using a dropper take 1 mL of saliva solution and add it to the three test tubes.
7. Immediately, transfer 2 drops from test tube A to the first series of test tubes (triplicate) kept on the stand marked as pH 4. Similarly, transfer 2 drops from test tube B to the first series of test tubes (triplicate) kept on the stand marked as pH 6.8 and transfer 2 drops from test tube C to the first series of test tubes (triplicate) kept on the stand marked as pH 9. Note the time as time 0 and observe the color of the iodine solution (it turns blue).
8. After 2 min, transfer 2 drops from test tube A to the second series of test tubes (triplicate) kept on the stand marked as pH 4. Do the same for test tubes B and C. Note the time as time 0 and observe the color of the iodine solution.
9. Repeat the procedure after every 2 min, till the color of the iodine solution does not change and note the time.

For temperature
1. Take three sets of test tubes with eight tubes in each set. Mark the test tube stands at 0°C, 37°C, and 50°C. In each test, take 0.5 mL of iodine solution (orange-brown).
2. Take three separate test tubes and in one beaker take 15 mL of 1% starch solution, 3 mL of 1% NaCl solution, and 3 mL of pH 6.8 buffer and divide it into the three test tubes equally.
3. Put one test tube in the ice and mark it as A; put one test tube at 37°C and mark it as B; put one test tube at 50°C and mark it as C.
4. In a beaker, collect saliva and using a dropper take 1 mL of saliva solution, and add it to the three test tubes.
5. Immediately, transfer 2 drops from test tube A to the first series of test tubes (triplicate) kept on the stand marked as 0°C. Similarly, transfer 2 drops from test tube B to the first series of test tubes (triplicate) kept on the stand marked as 37°C and transfer 2 drops from test tube C to the first series of test tubes (triplicate) kept on the stand marked as 50°C. Note the time as time 0 and observe the color of the iodine solution (it turns blue).
6. After 2 min, transfer 2 drops from test tube A to the second series of test tubes (triplicate) kept on the stand marked as 0°C. Do the same for test tubes B and C. Note the time as time 0 and observe the color of the iodine solution.
7. Repeat the procedure after every 2 min, till the color of the iodine solution does not change and note the time.

Analysis and statistics

The iodine solution of pH 6.8 would reach achromic point fastest at 37°C.

Precursor techniques
1. 1% Starch solution—Add 1 g of starch to make a solution of 100 mL with water. Heat it to 90°C to dissolve it.
2. 1% NaCl solution—Add 1 g of NaCl to make a solution of 100 mL with water.
3. 1% Iodine solution—Add 1 g of iodine to make a solution of 100 mL with water.

Safety considerations and standards
1. The test is sensitive to the temperature and pH, so make sure both of these are exactly as described in the experiment.

Pros and cons

Pros	Cons
Very easy method and requirements are minimal	Not a quantitative method

Summary
1. The enzyme amylase is used to degrade the starch at appropriate temp. and pH.
2. The starch that has been degraded will not be able to change the color of the iodine solution.
3. NaCl increases the rate of reaction, thus aiding the hydrolysis of starch.

Definition
Determination of the activity of amylase enzyme (crude) extracted from sweet potato.

Rationale
Enzyme amylase act on starch to produce reducing sugars that react with dinitrosalicylic acid to produce a brown-colored product. This product is estimated colorimeter and the amount of reducing sugars produced by the action of the enzyme is determined by the standard graph.

Materials, equipment, and reagents
A. **Reagents**: DNS solution, 1% starch solution, 0.1 M sodium acetate buffer pH 4.7.
B. **Glassware**: Test tube, test tube holder, pipettes, test tube stand.
C. **Instrument**: Water bath.

Protocol
1. Take a medium-sized sweet potato, wash, peel, and cut it into small cubes.
2. Weigh 2 g of the cubes and blend it in a blender with 10 mL of 0.1-M sodium acetate buffer pH 4.7.
3. Keep it at 4°C for 24 h.
4. Filter the extract with cheesecloth.
5. Centrifuge it at 13,000 g for 10 min and transfer the supernatant to a fresh test tube, discard the pellet. This is a crude extract of the enzyme.

Preparation of standard curve:
(A) **Stock standard (1 mg/mL of glucose)**: Weigh 100 mg of glucose and add 100 mL of distilled water to it.
(B) **Working standard (0.25 mg/mL of glucose)**: Dilute 10 mL of stock solution to a final volume of 40 mL by adding distilled water to it.

6. Take six test tubes and mark them as 1–6. Pipette standard glucose solution to the test tubes and then add distilled water to it to make up the final volume 2 mL.
7. Take three more test tubes and label them as 7, 8, and 9. Add 1 mL of starch solution to all three of them.
8. Add 1, 0.5, and 0 mL of sodium acetate buffer to 7, 8, and 9 test tubes, respectively.
9. Add 0, 0.5, and 1 mL of the diluted enzyme to 7, 8, and 9 test tubes, respectively.
10. Mix the contents of the test tube and incubate them for 15 min at 37°C.
11. Stop the enzymatic reaction by adding 1 mL of DNS solution to all the test tubes (1–9).
12. Put all the tubes in a boiling water bath for 5 min.
13. Cool the test tubes to room temperature.
14. Measure the optical density at 540 nm.

Standard curve:

Test Tube No.	Blank	1	2	3	4	5
Working solution (mL)	0	0.2	0.4	0.6	0.8	1.0
Water added (mL)	1	0.8	0.6	0.4	0.2	0
DNS reagent	1	1	1	1	1	1
Incubation in boiling water bath for 5 min						
OD at 540 nm	A_0	A_1	A_2	A_3	A_4	A_5
Final OD	A_0	$A_1 - A_0$	$A_2 - A_0$	$A_3 - A_0$	$A_4 - A_0$	$A_5 - A_0$

Enzymatic activity:

Test Tube No.	7	8	9
Vol. of the substrate (mL)	1	1	1
Vol. of buffer (mL)	1	0.5	0
Vol. of the enzyme (mL)	0	0.5	1
Incubation at 37°C for 15 min			
Vol. of DNS	1	1	1
Incubation in boiling water bath for 5 min			
OD at 540 nm	$T1$	$T2$	$T3$
Final OD	$T1 - A_0$	$T2 - A_0$	$T3 - A_0$

Analysis and statistics

Calculate the concentration of the test sample corresponding to the optical density obtained, using the linear standard graph.

Calculation

- Calculate the amount of reducing sugars formed by the activity of the enzyme. Then, the enzyme activity can be represented as follows:
- Enzyme activity =.................mg/15 min/mL of the diluted enzyme (obtained from the graph).
- Divide the above value by 15, it will be enzyme activity =.................mg/min/mL of diluted enzyme.
- For undiluted enzyme multiply the dilution factor of the enzyme, i.e., 25.

  Enzyme activity for undiluted enzyme = (..................... × 25) mg/min/mL of undiluted enzyme.

  Suppose total volume of enzyme extract = X mL which is obtained from Y mg of source.

  Then total enzyme activity is (.....................× 25) mg/min/mL multiplied by X/Y.

Precursor techniques

1. 0.1 M sodium acetate buffer (pH 4.7): Dissolve 3.4 g of sodium acetate in 200 mL of water and adjust the pH to 4.7 by adding glacial acetic acid and make the volume up to 250 mL.
2. 1% starch solution: Dissolve 1 g of starch to make a solution of 100 mL with water and heat up to 50°C to dissolve it.
3. Sodium potassium tartrate: Add 30 g of the salt to 100 mL of the final solution in water.
4. 3,5-Dinitrosalicylic acid: Dissolve 5 g of this reagent in 100 mL of water.
5. Dinitrosalicylic acid (DNS) reagent: Add 50 mL of (3) to 20 mL of (4) and make the solution to 1 L with water.
6. Enzyme dilution: Dilute crude enzyme in the ratio 1:25.

Safety considerations and standards

1. The test is sensitive to the temperature, so, cool down all the samples to room temperature before reading.
2. Always keep the enzyme extract at 4°C.

3. The dilution of the enzyme can be adjusted according to the color produced, but always remember the dilution factor while calculating the activity.

Pros and cons

Pros	Cons
Very easy method and requirements are minimal	

Summary

1. The enzyme amylase is extracted from sweet potato by grind it with buffer followed by centrifugation.
2. The activity of an enzyme is determined by the hydrolysis of starch to reducing sugars and the amount of reducing sugars is determined by the DNS method.

Definition

Determination of the specific activity of the amylase enzyme.

Rationale

The specific activity of an enzyme is the activity of the enzyme per milligram of total protein. It gives the measurement of enzyme purity. It is calculated by the amount of protein present in 1 mg in 1 mL of enzyme solution.

Materials, equipment, and reagents

A. **Reagents**: Lowry's reagent, Folin's reagent, BSA standards solution.
B. **Glassware**: Test tube, test tube holder, pipettes, test tube stand.
C. **Instrument**: Water bath.

Protocol

1. Take six test tubes. Label it as 1–5 and one blank.
2. Take 0.2–1 mL of working BSA solution in test tubes labeled as 1–6 and in blank do not add any BSA.
3. Add water to each test tube to make a final volume of 1 mL.
4. Take another test tube and add 1 mL of a diluted enzyme in it.
5. To each test tube, add 5 mL of Lowry's reagent and incubate it for 15 min in dark.
6. Add 0.5 mL of FC reagent to all test tubes.
7. Incubate them in dark at RT for 30 min.

Vol. of BSA (mL)	Vol. of Water (mL)	Conc. of Protein (µg)	Lowery's Reagent	Incubation for 5 min	Follin's Reagent	Incubation for 30 min	OD at 750 nm
0	1	0	5		0.5		
0.2	0.8	20	5		0.5		
0.4	0.6	40	5		0.5		
0.6	0.4	60	5		0.5		
0.8	0.2	80	5		0.5		
1	0	100	5		0.5		
1 mL enzyme solution	0	Unknown	5		0.5		

8. Measure the OD and calculate the concentration of protein in 1 mL of the enzyme.

Analysis and statistics

Calculate the concentration of protein in the enzyme sample corresponding to the optical density obtained, using the linear standard graph.

Calculation

The enzyme was diluted by 25 times in the previous experiment so multiply the concentration obtained by the graph.

The specific activity of the enzyme is = activity of enzyme/conc. of protein. (Here the activity of the enzyme is obtained from the previous experiment.)

Precursor techniques

Stock solution
1. BSA standard solution (1 mg/mL): weigh 10 mg of BSA and add it to water to make a final volume of 10 mL.
   Working solution
   Take 1 mL of the above solution and add 9 mL water to make a solution of conc. 100 μg/mL.
2. Lowery's solution:
   Solution A—add 2.85 g of sodium hydroxide and 14.3 g of sodium carbonate to water to make a solution of 500 mL.
   Solution B—add 1.423 g of copper sulfate pentahydrate in a final solution of 100 mL.
   Solution C—2.85 g of disodium tartrate dihydrate in 100 mL water.
   Mix the solution in a final ratio of 100:1:1.
3. Folin's reagent:
   5 mL of 2 N Folin and Ciocalteu's Phenol reagent and 6 mL of water.
   Prepare this solution fresh. It is a light-sensitive reagent.

Safety considerations and standards

1. The test is sensitive to the light, as Folin's reagent is light sensitive so the reaction should be carried out in dark.
2. Always keep the enzyme extract at 4°C and prepare the reagents fresh.

Pros and cons

Pros	Cons
Very easy method and requirements are minimal	

Summary

1. The concentration of protein is calculated by the Folin-Lowery method.
2. The activity of the enzyme is divided by conc. of protein to obtain the specific activity of the enzyme.

Definition

To determine V_{max} and K_m of the amylase enzyme.

Rationale

The extract amylase is used to carry out the hydrolysis of starch and by DNA method of estimation of reducing sugar produced, the activity of the enzyme is calculated, and thereby the enzyme kinetics can be obtained. Details about the Michaelis-Menten constant and other parameters have been discussed earlier.

Materials, equipment, and reagents

A. **Reagents**: DNS solution, 1% starch solution, 0.1-M sodium acetate buffer pH 4.7. The diluted enzyme (crude).
B. **Glassware**: Test tube, test tube holder, pipettes, test tube stand.
C. **Instrument**: Water bath.

Protocol

1. Take 10 test tubes in two sets. Mark test tubes of one set as C1 to C10 and another set of test tubes as T1 to T10 (C stands for control, i.e., without enzyme and T stands for test sample).
2. Add 0.5 mL of the diluted enzyme (1:5) to each test tube in the set of tubes marked as T.
3. Add substrate (starch, in this case) in a range of 0.1–1 mL to various test tubes as indicated in the table.
4. Add buffer to the test tubes to make a final volume of 2 mL.
5. Mix the contents and incubate them at RT for 15 min.
6. Stop the reaction between starch and enzyme amylase by adding 1 mL of DNS solution to all the tubes.
7. Keep all the test tubes in a boiling water bath for 5 min.
8. Cool the test tubes and read the absorbance at 540 nm.
9. Calculate the activity of each test tube as mentioned in the previous experiment.
10. Plot the graphs accordingly.

Precursor techniques

1. 0.1-M Sodium acetate buffer (pH 4.7): Dissolve 3.4 g of sodium acetate in 200 mL of water and adjust the pH to 4.7 by adding glacial acetic acid and make the volume up to 250 mL.
2. 1% starch solution: Dissolve 1 g of starch to make a solution of 100 mL with water and heat up to 50°C to dissolve it.

3. Sodium potassium tartrate: Add 30 g of the salt to 100 mL of the final solution in water.
4. 3,5-Dinitrosalicylic acid: Dissolve 5 g of this reagent in 100 mL of water.

5. Dinitrosalicylic acid (DNS) reagent: Add 50 mL of (3) to 20 mL of (4) and make the solution to 1 L with water.
6. Enzyme dilution: Dilute crude enzyme in the ratio 1:25.

Test Tube	Vol. of the Enzyme (mL)	Vol. of the Substrate (mL)	Vol. of Buffer (mL)	At RT for 15 min	Vol. of DNS (mL)	At Boiling Water Bath for 5 min	OD at 540 nm	Activity (ϑ_0)	$1/\vartheta_0$	S	1/S
C1	–	0.1	1.9		1						
T1	0.5	0.1	1.4								
C2	–	0.2	1.8		1						
T2	0.5	0.2	1.3								
C3	–	0.3	1.7		1						
T3	0.5	0.3	1.2								
C4	–	0.4	1.6		1						
T4	0.5	0.4	1.1								
C5	–	0.5	1.5		1						
T5	0.5	0.5	1.0								
C6	–	0.6	1.4		1						
T6	0.5	0.6	0.9								
C7	–	0.7	1.3		1						
T7	0.5	0.7	0.8								
C8	–	0.8	1.2		1						
T8	0.5	0.8	0.7								
C9	–	0.9	1.1		1						
T9	0.5	0.9	0.6								
C10	–	1.0	1.0		1						
T10	0.5	1.0	0.5								

Safety considerations and standards
1. The test is sensitive to the temperature, so, cool down all the samples to room temperature before reading.
2. Always keep the enzyme extract at 4°C.
3. The dilution of the enzyme can be adjusted according to the color produced, but always remember the dilution factor while calculating the activity.

Pros and cons

Pros	Cons
1. Very easy method and requirements are minimal	Not a quantitative method

Summary
1. The enzyme amylase is used to degrade the starch at appropriate temp. and pH.
2. The degradation of starch is indicated by the presence of reducing sugars, measured by the DNS method.
3. The activity of the enzyme is measured for the various concentration of starch and graphs are plotted.

CHAPTER 8

Nucleic Acid

1. Determination of the amount of DNA present in the solution by diphenylamine (DPA) method.
2. Determination of the amount of RNA present in the solution by orcinol method.
3. Isolation of plasmid DNA using the alkaline lysis method.
4. Preparation of plasmid DNA using boiling lysis method.
5. Plasmid DNA purification using polyethylene glycol.
6. Isolation of genomic DNA of *the unknown bacterial strain*.
7. Isolation of DNA from plant tissue.
8. Isolation of DNA from a eukaryotic cell using proteinase K and phenol.
9. Isolation of DNA from a eukaryotic cell.
10. Separation of bands of DNA by agarose gel electrophoresis.
11. Southern transfer and hybridization: to perform Southern blotting of the genomic DNA.
12. Isolation of RNA from animal cells.
13. Northern blotting.

Definition

Determination of the amount of DNA present in the solution by diphenylamine (DPA) method.

Rationale

All the deoxypentose give this reaction. In the presence of an acid, DNA gets depurinated, followed by dehydration of sugars to ω-hydroxylevulinylaldehyde. In acidic medium, this aldehyde condenses with diphenylamine and produces a deep blue-colored product. The colored product can be quantitated at 595 nm.

$$\text{Deoxypentose sugar} \xrightarrow{\text{Acidic medium}} \omega$$
$$-\text{Hydroxylevulinylaldehyde} \xrightarrow{\text{diphenylamine}} \text{Blue color}$$

Materials, equipment, and reagents

A. **Reagents**: DPA reagent, saline citrate buffer.
B. **Glassware**: Test tube, test tube holder, pipettes, cuvettes.
C. **Instrument**: Spectrophotometer.

Protocol

Preparation of standard curve:

(A) **Stock standard (10 mg/mL of DNA)**:
   Weigh 50 mg of DNA and add 5 mL of buffered saline to it.
   1. Take seven test tubes and label, including one blank.
   2. To these add 5–30 μL of the stock solution according to the following table.
   3. Simultaneously, take 200 μL of the unknown solution to it.
   4. Add 3 mL of DPA reagent to each test tube, mix well, and cover each of the test tubes.
   5. Put the tubes in a boiling water bath for 5 min.
   6. Cool the test tubes to room temperature.
   7. Measure the optical density at 595 nm.
   8. First, measure the OD of blank and make it zero.
   9. Then take the OD of the rest of the test tubes. Wash the cuvettes properly after taking the value of each sample.

The standard calibration curve with concentration plotted on the X-axis and optical density on the Y-axis:

Test Tube No.	Blank	1	2	3	4	5	6	Test Sample
Stock solution (μL)	0	5	10	15	20	25	30	200
Water added (μL)	200	195	190	185	180	175	170	0
DPA reagent	3	3	3	3	3	3	3	3
Heat in boiling water bath for 5 min. Cool it to room temperature								
OD at 595 nm	A_0	A_1	A_2	A_3	A_4	A_5	A_6	A_T

Analysis and statistics

Calculate the concentration of the test sample corresponding to the optical density obtained, using the linear standard graph.

Precursor techniques

1. DPA solution: Dissolve 1 g of diphenylamine in 100 mL of glacial acetic acid and 2.5 mL of conc. sulfuric acid.
2. Buffered saline: Mix 0.5 M of NaCl and 0.015-M sodium citrate and adjust the pH to 7.

Protocols in Biochemistry and Clinical Biochemistry. https://doi.org/10.1016/B978-0-12-822007-8.00010-6

Calculation

The concentration can be obtained using a standard graph. The dilution (if made) of the test sample should be multiplied to obtain the final concentration of the test sample solution. In this case, the concentration of DNA obtained is for 200 µL, so, multiply it by 5 to get the final concentration per mL of solution.

Safety considerations and standards

1. The test is sensitive to the temperature, so, cool down all the samples to room temperature before reading.
2. Glacial acetic acid and sulfuric acid can cause severe burns, so, it should be handled carefully.

Pros and cons

Pros	Cons
1. An easy and quick method for determining the amount of DNA 2. This method can be used for relatively crude extracts, where it is not possible to directly measure the UV absorbance of denatured DNA	Not suitable for the determination of concentration in a complex mixture of sugars

Alternative methods and protocols

Using ethidium bromide.

Troubleshooting

Problem	
The value of unknown is more than the most concentrated standard	Dilute the unknown and again read the absorbance
The value of the standard is lower than that of the blank	Make sure the amount of DNA added is correct and the wavelength for measuring absorbance is correct

Summary

1. The method utilizes colored compound with absorbance at 595 nm to measure the amount of DNA.
2. In an acidic medium in the presence of diphenylamine, the DNA gives a colored product that can be quantified using a spectrophotometer and the amount of DNA can be calculated.

Definition

Determination of the amount of RNA present in the solution by orcinol method.

Rationale

When pentose is heated with concentrated hydrochloric acid, it forms furfural. Orcinol reacts with this furfural to give a green-colored compound. Ferric chloride acts as a catalyst. Only purine nucleotides give a significant reading.

Materials, equipment, and reagents

A. **Reagents**: Orcinol reagent, RNA solution.
B. **Glassware**: Test tube, test tube holder, pipettes, cuvettes.
C. **Instrument**: Spectrophotometer.

Protocol

1. Take 2 mL of the nucleic acid solution to add 3 mL of orcinol to it.
2. Heat the solution for 20 min in a boiling water bath.
3. Let it cool down and read the absorbance at 665 nm against blank containing no nucleic acid.

Analysis and statistics

Calculate the concentration of the test sample corresponding to the dilution done for the sample.

Precursor techniques

1. Orcinol solution: Dissolve 0.025 g of ferric chloride ($FeCl_3 \bullet 6H_2O$) in 25 L of concentrated HCl to this add 875 mL of 6% orcinol in ethanol (freshly prepared).

Safety considerations and standards

1. The test is sensitive to the temperature, so, cool down all the samples to room temperature before reading.
2. Sulfuric acid can cause severe burns, so, it should be handled carefully.

Pros and cons

Pros	Cons
1. An easy and quick method for determining the amount of RNA 2. This method can be used for relatively crude extracts, where it is not possible to directly measure the UV absorbance of denatured DNA	Not suitable for the determination of concentration in a complex mixture of sugars

Alternative methods and protocols

Using ethidium bromide.

Troubleshooting

Problem

The value of unknown is more than the most concentrated standard	Dilute the unknown and again read the absorbance
The value of the standard is lower than that of the blank	Make sure the amount of RNA added is correct and wavelength for measuring absorbance is correct
	Prepare the orcinol reagent freshly

Summary

1. The method utilizes the formation of a colored compound with absorbance at 665 nm to measure the amount of RNA.
2. In an acidic medium in the presence of orcinol, the RNA gives a colored product that can be quantified using a spectrophotometer and the amount of RNA can be calculated.

Definition

Isolation of plasmid DNA using the alkaline lysis method.

Rationale

In 1979, Birnboim and Doly discovered that when the bacterial suspension is exposed to a strong anionic detergent at a high pH, the cell wall of bacteria opens, the chromosomal DNA, and proteins get denatured, and the plasmid DNA is released. At high pH, the DNA (chromosomal and plasmid) gets denatured and their strands get separated due to breaking of a hydrogen bond between the bases of adjacent strands; however, the closed circular plasmid DNA strands do not get separated as the strands are intertwined. The denaturation occurs as the alkaline solution is rich in hydroxide ions $(OH)^-$ and they pull the hydrogen ions of the bases and disrupt the hydrogen bond. If the exposure to alkaline pH is not for a long period, the two strands of plasmid DNA can again get renatured when the pH of the solution gets neutral. During lysis, the broken cell wall, bacterial proteins, and denatured chromosomal DNA assemble in large complexes coated with dodecyl sulfate. During neutralization, when the sodium ions are replaced by potassium ions, these large complexes can be precipitated and removed by centrifugation. The supernatant contains plasmid DNA that can be recovered by various methods like centrifugation using sodium chloride.

Materials, equipment, and reagents

A. **Reagents**: Alkaline lysis solution I, II, and III, antibiotics, ethanol, phenol-chloroform solution, TE buffer (pH 8) containing RNase A.
B. **Glassware**: Conical flask, cotton plug, spirit lamp, centrifuge tubes.
C. **Instrument**: Tabletop centrifuge.

Protocol

In the protocol, mini, midi, and maxi preparation of plasmid are used for small, medium, and large-scale production of plasmid, respectively.

1. Take 2 mL of LB medium or terrific broth, containing appropriate antibiotic (for kanamycin, final concentration should be 50 μg/mL and for ampicillin 100 μg/mL). Inoculate it with a single colony of bacteria and incubate it overnight at 37°C, 220 rpm in a shaker cum incubator.
2. The next morning, take 1.5 mL of the culture in a centrifuge tube and centrifuge it at max speed for 30 s at 4°C. (For midi preparation, take 10 mL of culture and centrifuge it at 2000g for 10 min at 4°C; for maxi preparation, take 500 mL of culture and centrifuge it at 2700g for 15 min.)
3. Remove the supernatant (medium) by aspiration, or the tube can be tilted carefully on a tissue paper without disturbing the pellet. Aspiration can be done by Pasture pipette attached to a vacuum line or simple using a pipette. If the medium is not removed properly from the pellet then the cell wall components will be retained in the medium which can inhibit the action of many restriction enzymes and the plasmid obtained will be resistant to cleavage by them. Alternatively, the bacterial pellet can be resuspended in an ice-cold STE buffer followed by centrifugation. The volume of STE buffer should be 25% the original culture volume for mini and midi preparation and for maxi preparation, 200 mL of STE should be used.
4. Resuspend the bacterial pellet obtained after centrifugation, in 100 μL of ice-cold alkaline lysis solution I by vigorous mixing so that the pellet is completely dispersed in it. (For midi preparation, use 200 μL of solution I; for maxi preparation use 18 mL of solution I with 2 mL of 10 mg/mL of lysozyme.)
5. Add 200 μL of alkaline lysis solution II, close it and mix the contents gently by inverting the tubes quickly. The whole surface inside the tube should come in contact with the solution. Store the tubes in ice for 2 min. (For midi preparation, use 400 μL

of solution II and maxi preparation 5 mL of the same for 5 min.)

6. Add 150 μL of ice-cold alkaline lysis solution III, close the tube and invert the tubes several times so that the contents of the tubes get mixed properly. Store the tubes in ice for 3–5 min. (For midi preparation, use 300 μL of solution III and maxi preparation 20 mL of the same for 10 min.)

7. Centrifuge the lysate of bacteria for 5 min at max speed and 4°C. Transfer the supernatant containing DNA in another centrifuge tube. Discard the pellet. (For maxi preparation, centrifuge at 20,000g for 30 min.)

8. Add an equal amount of phenol-chloroform-isoamyl alcohol mixture in the ratio of 25:24:1; mix it by vigorously shaking the tubes and centrifuge at max speed for 3 min at 4°C. Transfer the upper layer (aqueous layer) to a fresh centrifuge tube. In this process, as the phenol-chloroform is immiscible in water, two distinct phases will form after centrifugation. Phenol-chloroform layer comprises the nonpolar substances as lipids while the aqueous phase contains DNA, RNA, and soluble salts. The protein gets denatured on the addition of phenol and due to the presence of both hydrophobic and hydrophilic side chains present in them; it remains at the interface of the upper and lower phase. The aqueous phase rises to the top as the density of phenol-chloroform is higher. The chloroform increases the density of the organic phase. Isoamyl alcohol reduces the foaming between the interphase.

9. To the upper phase obtained from above add an equal volume of chloroform, mix it vigorously and centrifuge at max speed for 3 min at 4°C. Again, take the upper aqueous solution in a fresh tube. Phenol is slightly soluble in water but can be shifted to the organic phase by the addition of chloroform. Chloroform, in this step, will remove the residual phenol from aqueous solution.

10. Add 2 volumes of ice-cold ethanol solution and NaCl solution (final concentration of NaCl should be 0.2 M) to the above tube containing aqueous solution. The electrostatic attraction between the positive and negative charged ions Na^+ (of NaCl) and PO^{3-} (of DNA) is affected by the dielectric constant of a solution. The dielectric constant of the water is quite high; therefore, it is much difficult for both the charges to come close in the water. But when it is replaced by the ethanol the ions get close, the charge of DNA gets neutralized and the DNA gets precipitated.

11. Mix above and let it stand for 5 min at RT.

12. Centrifuge at max speed for 10 min at 4°C. (For maxi preparation, centrifuge for 15 min.)

13. Remove the supernatant and wash the pellet with 500 μL of 70% ethanol. A total of 70% of ethanol can wash the residual salt that gets precipitated with DNA. (For maxi preparation, use 1 mL of 70% ethanol.)

14. Remove the supernatant by aspiration and use disposable pipettes or tissue paper to remove the fluids adhered to the sides of the tubes.

15. Let the tube air dry till the ethanol gets evaporated. It is not advised to dry it in a desiccator or under vacuum, as the pellet then obtained becomes very difficult to dissolve.

16. Dissolve the DNA (pellet) in 50 μL of TE buffer (pH 8) containing 20 μg/mL DNase free RNase A. Tap the tube or vortex till the DNA gets dissolved. Store the solution in −20°C. (For midi preparation, dissolve it in 100 μL of TE buffer and for maxi preparation 3 mL of the same.)

Analysis and statistics
The quality of DNA can be checked by electrophoresis and it can be quantified using a spectrophotometer.

Precursor techniques
1. Alkaline lysis solution I: 50 mM glucose and 25 mM Tris-Cl (pH 8) 10 mM EDTA.
   Stock solutions:
   0.5 M glucose: Add 9 g of glucose to a final solution of 100 mL with water.
   1 M Tris-Cl (pH 8): Add 12.1 g of Tris base to 80 mL of water and adjust its pH using conc. HCl and make up the volume to 100 mL.
   1 M EDTA (pH 8): Add 29.24 g of EDTA to 80 mL water and adjust the pH to 8 using NaOH pellets and make up the volume to 100 mL.
   Now, take 10 mL of 0.5 M glucose, 2.5 mL of 1 M Tris-Cl, and 1 mL of 1 M EDTA mix all and make up the final volume to 100 mL with water.
   Autoclave it and keep at 4°C.
2. Alkaline solution II: 0.2 N NaOH and 1% SDS (freshly prepared).
   Stock solutions:
   10 N NaOH: Add 40 g of NaOH to a final volume of 100 mL in water.
   10% SDS: 10 g of SDS to a final volume of 100 mL in water.
   Take, 2 mL of 10 N NaOH and 10 mL of 10% SDS, make up the volume to 100 mL.

3. Alkaline lysis solution III.
   Stock solutions:
      5-M potassium acetate: Add 49 g of potassium acetate to 100 mL of the final solution with water.
      Add 60 mL of 5-M potassium acetate and 11.5 mL of glacial acetic acid and make up the final volume of 100 mL with water. Autoclave it and store it at 4°C.
4. Phenol-chloroform-isoamyl solution: Add 1 mL of isoamyl alcohol to 24 mL of chloroform. Mix an equal amount of equilibrated phenol and above solution of chloroform and isoamyl alcohol.
5. 10 × TE buffer: Mix 100 mM Tris-Cl and 10 mM EDTA (weigh 1.21 g of Tris-base, add it to 80 mL water and adjust its pH to 8 by adding conc. HCl, and make up the volume to 100 mL, this is 1-M Tris-Cl solution; weigh 29.24 g of EDTA, add it to 80 mL water, adjust its pH to 8 using NaOH pellets, and make up the volume to 100 mL, this is 1 M EDTA. To prepare 10 × TE buffer, take 10 mL of Tris-Cl and 1 mL of EDTA and make up the solution to 100 mL. For experiment, dilute 10 × TE buffer to 1 × by adding 10 mL of buffer to 90 mL of water.

Safety considerations and standards

1. The culture should be well aerated, to ensure it the volume of culture tube or flask should be four times greater than the volume of bacterial culture, the tube or flask should be capped loosely and the culture should be vigorously agitated during incubation.
2. Phenol and NaOH glacial acetic acid should be handled carefully.
3. Solution II should be prepared freshly. The treatment with solution II should not be prolonged.
4. The pellet obtained after ethanol precipitation is quite difficult to observe so it is advised to arrange the tubes in a centrifuge in a particular way, e.g., all tubes can be kept so that their plastic hinge points outwards, the pellet obtained will be at the farthest from the center of rotation, and it will be easy to locate it.

Pros and cons

Pros	Cons
Small- to large-scale plasmid preparation can be done	Relatively time-consuming
Suitable for small as well as large-sized plasmid	

Alternative methods
Boiling lysis method.

Summary
1. In this method of plasmid isolation, the cells are disrupted by NaOH, which provide high pH, and SDS.

2. It utilizes the property of plasmid DNA, which are intertwined, to get renatured on the introduction of high pH only for a short period, while the bacterial genomic DNA cannot renature.
3. This is followed by the extraction of DNA in aqueous phase with the treatment of phenol-chloroform which left the proteins and other debris.
4. Lastly, the ethanol precipitates out the DNA from aqueous solution.

Definition
Preparation of plasmid DNA using boiling lysis method.

Rationale
The prototype of this method was carried out by Holmes and Quigley in 1981. In the boiling lysis method, the bacterial suspension is subjected to a buffer containing lysozyme and TritonX-100, followed by heating at 100°C. Lysozyme attacks the proteoglycan cell wall of bacteria, while TritonX-100 is a mild nonionic detergent which acts as a surfactant and disrupts the lipid-lipid and lipid-protein interaction and does not denature the proteins. The heat cracks open the bacterial outer shell and also disrupt the hydrogen bond between the bases of DNA. Heat denatures the proteins and DNA. As the strands of DNA in a plasmid are intertwined, the plasmid DNA does not get separated from each other and again renatures when the temperature is lowered. The denatured chromosomal DNA and proteins get separated by centrifugation. The method is suitable for plasmids smaller than 15 kb in size and most strains of *E. coli*; however, it is not recommended for the strains that release a huge amount of carbohydrates when they are subjected to heat, detergent, and lysozyme. It is difficult to remove the carbohydrates from the plasmid preparations and it hampers the action of many restriction enzymes and polymerases. This method is also not suitable for the *E. coli* that express endonuclease A. The enzyme endonuclease A cannot be inactivated during boiling and the plasmid can be degraded when incubated with Mg^{2+}.

Materials, equipment, and reagents
A. **Reagents**: LB broth, lysozyme (10 mg/mL), antibiotics, ethanol, isopropanol, sodium acetate, STET, TE buffer (pH 8) containing RNase A.
B. **Glassware**: Conical flask, cotton plug, spirit lamp, centrifuge tubes.
C. **Instruments**: Boiling water bath, centrifuge.

Protocol
1. Take 2 mL of LB medium or terrific broth, containing appropriate antibiotic (for kanamycin final

concentration should be 50 µg/mL and for ampicillin 100 µg/mL). Inoculate it with a single colony of bacteria and incubate it overnight at 37°C, 220 rpm in a shaker cum incubator.

2. Next morning, take 1.5 mL of the culture in a centrifuge tube and centrifuge it at max speed for 30 s at 4°C (for large-scale preparation take 500 mL of culture and centrifuge it at 2700g for 15 min).

3. Remove the supernatant (medium) by aspiration, or the tube can be tilted carefully on a tissue paper without disturbing the pellet. Aspiration can be done by Pasture pipette attached to a vacuum line or simple using a pipette. If the medium is not removed properly from the pellet, then the cell wall components will be retained in the medium which can inhibit the action of many restriction enzymes and the plasmid obtained will be resistant to cleavage by them.

4. Resuspend the bacterial pellet obtained after centrifugation, in 350 µL of STET (for large-scale preparation, add 10 mL STET, and transfer it to Erlenmeyer flask).

5. To this, add 25 µL of lysozyme solution. The solution should be freshly prepared. Mix the contents gently by inverting the tubes 10 times or vortexing for 3 s (for large-scale preparation add 1 mL lysozyme).

6. Keep the tubes in a boiling water bath for 40 s. If the duration of heat treatment is increased, it will result in permanent denaturation of the DNA. The resulting cyclic coiled DNA is resistant to cleavage by restriction enzymes and has about twice the rate of migration as compared to that of superhelical DNA (for large-scale preparation, hold the Erlenmeyer flask using a clamp and hold it above the flame of Bunsen burner till the liquid starts boiling. Shake the flask constantly during boiling. Immediately immerse the bottom half of the flask in a 2-L beaker containing boiling water. Hold the flask in it for 40 s. Allow the flask to cool in ice-cold water for 5 min and transfer all the contents in an ultracentrifuge tube. Centrifuge it at 150,000g for 30 min at 4°C).

7. Centrifuge the lysate at max speed for 15 min at RT. Transfer the supernatant to a fresh centrifuge tube.

8. Add 40 µL of 2.5-M sodium acetate (pH 5.2) and 420 µL of isopropanol to the supernatant, mix it by vortexing, and allow it to stand for 5 min at RT (for large-scale preparation, add 0.6 volume of isopropanol, mix it, and let it stand for 10 min at RT).

9. Centrifuge it at max speed for 10 min at 4°C. Before centrifugation arrange the tubes in a particular manner to easily locate the DNA (for large-scale preparation, centrifuge at 12,000g for 15 min).

10. Decant the supernatant and dissolve the pellet in 100 µL of TE buffer (for large-scale preparation, add dissolve the pellet in 3 mL of TE buffer, this can be followed by treatment with phenol-chloroform-isoamyl alcohol).

11. Now add an equal volume of a mixture of phenol-chloroform-isoamyl alcohol. The purpose of the addition of the mixture of phenol-chloroform-isoamyl alcohol has been discussed in the previous experiment.

12. Mix it vigorously and centrifuge it at max speed for 3 min. Transfer the upper layer (aqueous layer) to a fresh centrifuge tube.

13. To the upper phase obtained from above add an equal volume of chloroform, mix it vigorously, and centrifuge at max speed for 3 min at 4°C. Again, take the upper aqueous solution in a fresh tube. The purpose of the addition of chloroform has been discussed in the previous experiment (follow the same for large-scale preparation too).

14. Transfer the supernatant in a fresh tube and add 0.1 volume of 3-M sodium acetate (pH 5.2) and 2.5 volume of ethanol (for large-scale preparation, add 2 volume of ethanol, and let it incubate for 1 h in ice, keeping on ice is optional).

15. Centrifuge it at max speed for 10 min at 4°C (centrifuge at 12,000g for 20 min).

16. Aspirate the supernatant by an aspirator or remove it using pipettes. Remove the traces of liquid by inverting the tube and using tissue paper.

17. Wash the pellet with 1 mL of 70% ice-cold ethanol by centrifuging it at max speed for 10 min at 4°C (for large-scale preparation, add 2 mL of 70% ice-cold ethanol).

18. Remove the supernatant and air dry the tubes to remove the traces of ethanol. Dissolve the DNA (pellet) in 50 µL of TE buffer (pH 8) containing 20 µg/mL DNase-free RNase A. Tap the tube or vortex till the DNA gets dissolved. Store the solution in −20°C.

Analysis and statistics
The quality of DNA can be checked by electrophoresis, and it can be quantified using a spectrophotometer.

Precursor techniques
1. STET: 10 mM Tris-Cl (pH 8.0), 0.1 M NaCl, 1 mM EDTA (pH 8.0) 5% (v/v) Triton X-100
   Stock solutions:
   1 M Tris-Cl (pH 8): Add 12.1 g of Tris base to 80 mL of water and adjust its pH using conc. HCl and make up the volume to 100 mL.

1 M EDTA (pH 8): Add 29.24 g of EDTA to 80 mL water and adjust its pH to 8 using NaOH pellets and make up the volume to 100 mL.

1 M NaCl: Add 5.85 g of NaCl to 100 mL of water.

Now, take 1 mL of 1 M Tris-Cl, 100 μL of 1 M EDTA, 10 mL of NaCl to 95 mL of water mix evenly and now add 5 mL of Triton X-100 and mix it (detergent should be added last to avoid misreading of the measuring cylinder due to frothing).

2. 2.5-M Sodium acetate (pH 5.2): Weigh 2 g of sodium acetate and dissolve it in water to make 6 mL of water, adjust the pH using acetic acid, and make up the volume to 10 mL.

3. Phenol-chloroform-isoamyl solution: Add 1 mL of isoamyl alcohol to 24 mL of chloroform. Mix an equal amount of equilibrated phenol and above solution of chloroform and isoamyl alcohol.

4. 10× TE buffer: Mix 100 mM Tris-Cl and 10 mM EDTA (weigh 1.21 g of Tris-base, add it to 80 mL water and adjust its pH to 8 by adding conc. HCl and make up the volume to 100 mL, this is 1-M Tris-Cl solution; weigh 29.24 g of EDTA, add it to 80 mL water, adjust its pH to 8 using NaOH pellets and make up the volume to 100 mL, this is 1 M EDTA. To prepare 10× TE buffer, take 10 mL of Tris-Cl and 1 mL of EDTA, and make up the solution to 100 mL. For experiment, dilute 10× TE buffer to 1× by adding 10 mL of buffer to 90 mL of water.

Safety considerations and standards

1. The culture should be well aerated, and to ensure it, the volume of culture tube or flask should be four times greater than the volume of bacterial culture, the tube or flask should be capped loosely and the culture should be vigorously agitated during incubation.
2. The pellet obtained after ethanol precipitation is quite difficult to observe so it is advised to arrange the tubes in a centrifuge in a particular way, e.g., all tubes can be kept so that their plastic hinge points outwards, the pellet obtained will be at the farthest from the center of rotation and it will be easy to locate it.
3. The boiling should be strictly for 40 s.

Pros and cons

Pros	Cons
Relatively less no. of solutions is used in the process	Suitable for small plasmids only ≤15 kb
Less time consuming	Not suitable for strains of E. coil that produces a large number of carbohydrates
	Not suitable for E. coli expressing endonuclease A

Alternative methods

Alkaline lysis method and lysis with SDS method.

Summary

1. In this method, lysozyme attacks the cell wall of bacteria, while TritonX-100 disrupts the cell membrane.
2. This is followed by heat treatment which denatures the plasmid as well as genomic DNA, but after a short period, when the temp. is ice cold, the renaturation of plasmid DNA only occurs.
3. The DNA is extracted in aqueous solution by using phenol-chloroform extraction.
4. It is then precipitated by ethanol and collected after centrifugation.

Definition

Plasmid DNA purification using polyethylene glycol.

Rationale

The precipitation method was first developed by Richard Treisman and is widely used to purify the large scale of DNA obtained from the alkaline lysis method. In the protocol, the DNA is treated with lithium chloride which precipitates the large RNA, while small RNAs are digested by the treatment with RNase. LiCl is a strong dehydrating agent which lowers the solubility of RNA and removes proteins from the chromatin. Therefore, the high-molecular-weight RNA and protein can be removed leaving behind the DNA. The treatment of LiCl is followed by the treatment of PEG in high salt solution which precipitates large plasmid DNA, while the short DNA and RNA fragments retained in the solution. The precipitated plasmid DNA is purified by extraction with phenol-chloroform followed by ethanol precipitation. However, this method of separation is not efficient in separating the nicked circular plasmid from the covalently closed circular plasmid.

Materials, equipment, and reagents

A. **Reagents**: Crude plasmid preparation, ethanol, 3-M sodium acetate pH 5.2, 5 M LiCl, PEG-MgCl$_2$ solution, phenol-chloroform, TE buffer (pH 8) containing RNase A, TE buffer (pH 8).
B. **Glassware**: Centrifuge tubes.
C. **Instruments**: Ice water bath, centrifuge.

Protocol

1. Take a centrifuge tube and transfer 3 mL of large-scale crude plasmid preparation and keep it on a water-ice bath and chill it to 0°C.
2. To this, add 3 mL of an ice-cold 5 M LiCl and mix well. Centrifuge it 12,000*g* for 10 min at 4°C.

3. Transfer the supernatant to a fresh centrifuge tube and add an equal volume of isopropanol to it, mix well. The nucleic acids will get precipitated.
4. Centrifuge above at 12,000g for 10 min at RT. The nucleic acid will get pelleted.
5. Decant the supernatant and invert the tube, rinse the pellet, and tube with 70% ethanol at RT.
6. Decant the supernatant and invert the tube on a tissue paper to remove the last traces of ethanol.
7. Dissolve the damp nucleic acid pellet in 500 μL of TE (pH 8.0) containing RNase A. Let it stand at RT for 30 min.
8. Extract this mixture of plasmid and RNase with phenol-chloroform-isoamyl alcohol followed by chloroform-isoamyl alcohol, as mentioned in previous experiments.
9. Using standard ethanol precipitation, as mentioned in previous experiments, recover the DNA.
10. Dissolve the plasmid DNA, recovered as a pellet from the above step, in 1 mL autoclaved double distilled water. Add 0.5 mL of PEG-MgCl$_2$ solution to it.
11. Incubate the above solution for more than 10 min and then centrifuge it at max speed for 20 min at RT. Collect the precipitated plasmid DNA as a pellet.
12. Resuspend the pellet in 0.5 μL of 70% ethanol, to remove the traces of PEG and collect the nucleic acid by centrifuging it at max speed for 5 min at RT.
13. Aspirate the ethanol, repeat the above step followed by aspiration of 70% ethanol.
14. Air-dry the pellet for 20–25 min, to evaporate the ethanol.
15. Dissolve the pellet in 500 μL of TE (pH 8, without RNase).
16. Store the DNA at −20°C.

Analysis and statistics
The quality of DNA can be checked by electrophoresis and it can be quantified using a spectrophotometer.

Precursor techniques
1. 3-M sodium acetate pH 5.2: Add 24.6 g of sodium acetate to 70 mL of water and adjust the pH to 5.2 using acetic acid.
2. PEG-MgCl$_2$ solution: 40% polyethylene glycol (PEG 8000) and 30 mM MgCl$_2$. Sterilize the solution by passing through the filter.
3. Phenol-chloroform-isoamyl alcohol: As described in the previous experiment.
4. TE buffer (pH 8) containing RNase A: As described in the previous experiment.

Safety considerations and standards
1. Phenol, PEG, and sodium acetate should be handled carefully.
2. The pellet obtained after ethanol precipitation is quite difficult to observe so it is advised to arrange the tubes in a centrifuge in a particular way, e.g., all tubes can be kept so that their plastic hinge points outwards, the pellet obtained will be at the farthest from the center of rotation and it will be easy to locate it.

Pros and cons

Pros	Cons
Relatively, easy method of purification of plasmid	It cannot efficiently separate the nicked circular plasmids from the closed circular plasmid

Alternative methods
CsCl method of plasmid purification.

Summary
1. In this method, LiCl is used as a denaturant, which lowers the solubility of RNA and removes proteins from chromatin.
2. RNA is further degraded by RNase.
3. PEG precipitates the large plasmid DNA.
4. This is followed by phenol-chloroform extraction and ethanol precipitation.

Definition
Isolation of genomic DNA of *the unknown bacterial strain*.

Rationale
In 1979, the protocol for DNA isolation from bacteria was developed by Marmur. This method is a modified version of Marmur's method. In this method, the pellet of cells is resuspended in resuspension buffer with RNase A, which degrades RNA. This protocol comprises enzymes for both Gram-positive and Gram-negative bacteria. Achromopeptidase has bacteriolytic activity against the cell wall of Gram-positive bacteria, while lysozyme does the same against Gram-negative bacteria. SDS solubilizes the lipids and proteins of the cell membrane. Proteinase K acts on the proteins and digests them. As stated in the earlier protocol, the DNA will be extracted using phenol:chloroform:isoamyl alcohol solution and can be precipitated by ethanol.

Materials, equipment, and reagents
A. **Reagents**: Resuspension buffer with RNase A, ethanol, phenol-chloroform solution, TE buffer (pH 8) containing RNase A, Achromopeptidase (50 kU/mL), lysozyme (20,000 kU/mL), proteinase K (20 mg/mL).

B. **Glassware**: Conical flask, cotton plug, spirit lamp, centrifuge tubes.
C. **Instrument**: Tabletop centrifuge.

Protocol

1. Pick up a single colony from the agar plate and incubate it in 10 mL of LB medium.
2. Transfer 10 mL of a culture grown to mid-late log phase (0.5–0.7 at OD_{600}) to afresh falcon tube.
3. Centrifuge it at 7500 rpm for 10 min. Discard the supernatant.
4. Add 460 μL of RNase in buffer P1, resuspend it, and transfer it to a microfuge tube.
5. Add 8 μL lysozyme and 5 μL of achromopeptidase, mix it gently, and incubate it for 60 min at 37°C.
6. Add 30 μL of 10% SDS and 3 μL proteinase K, mix it gently, and then incubate it for 60 min at 50°C.
7. Add an equal amount of solution of phenol:chloroform:isoamyl alcohol, mix it gently by inverting the tubes for 10 min.
8. Centrifuge it at 12,000 rpm for 15 min.
9. Transfer the upper aqueous phase to another fresh microfuge tube.
10. Add chloroform:isoamyl alcohol solution to it (equal amount) and invert it several times gently for 10 min.
11. Centrifuge at 12,000 rpm for 15 min.
12. Take the upper layer in a fresh microfuge tube.
13. Add an equal volume of ice-cold ethanol and mix it gently by inversion.
14. Centrifuge it at 12,000 rpm for 20 min.
15. Carefully decant the supernatant and let the pellet air dry.
16. Dissolve the pellet in 50 μL of TE buffer and store it at −20°C.

Analysis and statistics

The quality of DNA can be checked by electrophoresis, take 5 μL of DNA, and run in 1.5% agarose gel. It can be quantified using a spectrophotometer.

Precursor techniques

1. Resuspension buffer: 50 mM Tris-Cl (pH 8) 10 mM EDTA. Add RNase A to a final concentration of 100 μg/mL.
   Stock solutions:
   1 M Tris-Cl (pH 8): Add 12.1 g of Tris base to 80 mL of water and adjust its pH using conc. HCl and make up the volume to 100 mL.
   1 M EDTA (pH 8): Add 29.24 g of EDTA to 80 mL water and adjust the pH to 8 using NaOH pellets and make up the volume to 100 mL.

Now, take 500 μL of 1 M Tris-Cl and 100 μL of 1-M EDTA mix all and make up the final volume to 10 mL with water.
Autoclave it and keep at 4°C.
2. 10% SDS: 10 g of SDS to a final volume of 100 mL in water.
3. 20 mg/mL proteinase K: Weigh 200 mg of proteinase K and dissolve it in 10 mL of water.
4. Phenol:chloroform:isoamyl alcohol: As described earlier.
5. TE buffer: As described earlier.

Safety considerations and standards

1. The culture should be well aerated, to ensure it the volume of culture tube or flask should be four times greater than the volume of bacterial culture, the tube or flask should be capped loosely, and the culture should be vigorously agitated during incubation.
2. Phenol and NaOH glacial acetic acid should be handled carefully.
3. The pellet obtained after ethanol precipitation is quite difficult to observe so it is advised to arrange the tubes in a centrifuge in a particular way, e.g., all tubes can be kept so that their plastic hinge points outward, the pellet obtained will be at the farthest from the center of rotation, and it will be easy to locate it.

Pros and cons

Pros	Cons
Small to large-scale genomic DNA preparation can be done	

Summary

1. In this method, lysozyme is used to degrade the cell wall of Gram-negative bacteria while achromopeptidase is effective against the cell wall of Gram-positive bacteria.
2. SDS solubilize the membrane and proteins are digested by proteinase K.
3. After this the phenol-chloroform extraction and ethanol precipitation.

Definition

Isolation of DNA from plant tissue.

Rationale

Isolation of DNA from plant tissue is comparatively difficult from that in animals. In different species of animals, the same type of tissues has similar characteristics, while among various species of plants the levels of metabolites and structural biomolecules differ greatly, especially, polysaccharides and polyphenols are quite problematic during extraction. Inclusion of CTAB in the process can aid in the separation of polysaccharides during purification, while reagents such as polyvinylpyrrolidone can help in the removal of polyphenols. Polysaccharides and DNA have different solubilities in CTAB, which in turn depends on the concentration of sodium chloride.

In high salt concentrations, polysaccharides are insoluble in CTAB, while at lower salt concentrations, DNA is insoluble. The DNA or polysaccharides can be precipitated by adjusting the salt concentration. Similarly, the interaction of DNA and phenolic rings can be prevented by the addition of polyvinylpyrrolidone.

Materials, equipment, and reagents

A. **Reagents**: Extraction buffer, RNase A, phenol, chloroform, isoamyl alcohol, ethanol, TE buffer (pH 8.0).
B. **Glassware/plasticware**: Mortar pestle, centrifuge tubes, microfuge tubes.
C. **Instrument**: Tabletop centrifuge.

Protocol

1. Weigh 0.3 g of plant tissue, clean it, and chop it off using a clean razor blade and grind the plant sample using liquid nitrogen, in a mortar and pestle, to a fine powder.
2. Weigh it and transfer it to a tube and add 500 μL of extraction buffer for per 0.1 g of homogenized tissue. Vortex the contents and place it in a water bath preset at 60°C for 30 min.
3. Centrifuge the contents at 13,000 rpm for 5 min.
4. Transfer the supernatant to a fresh tube and add 5 μL of RNase solution A and incubate for 20 min at 37°C.
5. Add an equal amount of phenol:chloroform:isoamyl alcohol, mix it properly, and centrifuge it at 13,000 rpm for 3 min. Transfer the upper phase to a fresh tube and repeat the process.
6. Take the upper layer to a fresh tube, add an equal amount of chloroform:isoamyl alcohol and mix it, and centrifuge it at 13,000 rpm for 3 min.
7. Take the upper layer to another fresh tube and add 2 volumes of ethanol (ice cold) and incubate at −20°C for 145 min.
8. Centrifuge it at 13,000 rpm for 10 min, discard the supernatant and wash pellet with 500 μL of ice-cold 70% ethanol, by centrifuging it for 5 min at 13,000 rpm.
9. Decant the ethanol and let the pellet air dry.
10. Dissolve the pellet in 50 μL of TE (pH 8.0).

Analysis and statistics

The quality of DNA can be checked by electrophoresis, take 5 μL of DNA, and run in 1.5% agarose gel. It can be quantified using a spectrophotometer.

Precursor techniques

1. Extraction buffer: 2% cetyl trimethylammonium bromide, 1% polyvinyl pyrrolidone, 100 mM Tris-HCl, 1.4 M NaCl, and 20 mM EDTA.
   Stock solutions:
      1 M Tris-Cl (pH 8): Add 12.1 g of Tris base to 80 mL of water and adjust its pH using conc. HCl, and make up the volume to 100 mL.
      1 M EDTA (pH 8): Add 29.24 g of EDTA to 80 mL water and adjust its pH to 8 using NaOH pellets and make up the volume to 100 mL.
      5 M NaCl: Weigh 29.25 g of NaCl and dissolve it in water to make a solution of 100 mL.
      Now, take 1 mL of 1 M Tris-Cl, 200 μL of 1 M EDTA, 2.8 mL of 5-M NaCl mix all, then add 0.2 g of cetyltrimethylammonium bromide and 0.1 g of polyvinyl pyrrolidone, and make up the final volume to 10 mL with water.
      Filter it and keep at 4°C.
2. Phenol:chloroform:isoamyl alcohol: As described earlier.
3. TE buffer: As described earlier.

Safety considerations and standards

1. Liquid nitrogen should be handled carefully as it can cause severe burns, the mortar and pestle should be cooled before adding liquid nitrogen to it. Liquid nitrogen should be added in batches.
2. The culture should be well aerated, to ensure it the volume of culture tube or flask should be four times greater than the volume of bacterial culture, the tube or flask should be capped loosely, and the culture should be vigorously agitated during incubation.
3. Phenol and NaOH glacial acetic acid should be handled carefully.
4. The pellet obtained after ethanol precipitation is quite difficult to observe so it is advised to arrange the tubes in a centrifuge in a particular way, e.g., all tubes can be kept so that their plastic hinge points outward, the pellet obtained will be at the farthest from the center of rotation, and it will be easy to locate it.

Pros and cons

Pros	Cons
The easiest way to isolate the DNA from plants	Liquid nitrogen is not easily available

Summary

1. In this process, the use of CTAB and pyrrolidone helps in the separation of DNA from polysaccharides and polyphenols, respectively.
2. The liquid nitrogen is used to disrupt the cell wall using mortar and pestle.
3. The phenol-chloroform solution is used to extract the DNA in aqueous solution.

Definition

Isolation of DNA from the eukaryotic cell using proteinase K and phenol.

Rationale

This method is a modified version of a method originally described by Daryl Stafford and colleagues. In this protocol, proteinase K is used which digest the proteins followed by the extraction of nucleic acid with phenol. The ammonium acetate is used to precipitate the proteins and ethanol precipitates DNA. The protocol is used when a large amount of DNA is required.

Materials, equipment, and reagents

A. **Reagents**: 10-M ammonium acetate, dialysis buffer, ethanol, lysis buffer, phenol, TE (pH 8.0), Tris buffer saline TBS, proteinase K (20 mg/mL), dialysis tubing clips, Shepherd's cooks, rocking platform, tube mixer or roller apparatus vacuum aspirator equipped with traps.
B. **Glassware/plasticware**: Cut off yellow tips, wide bore pipettes (0.3 cm orifice), tips, gel sealing tape, or gel caster.
C. **Instruments**: Sorvall centrifuge with H1000B and SS-34 rotors (or their equivalents), electrophoresis tank, spectrophotometer, water bath.

Protocol: Monolayers or suspensions of mammalian cells, or fresh tissue, or blood samples

From monolayer of cultured cells:

1.a. Take one batch of cultured dishes with cells grown approximately, more than 80% of confluency. Place it on an ice bucket. Remove the medium by aspiration and wash the monolayer of cells with ice-cold PBS. Aspirate it and again repeat the washing process quickly.
1.b. Scrape the cells using a rubber policeman in 1 mL of PBS. With the help of a pasture pipette transfer, the cell suspension to a centrifuge tube kept on ice. Process all the dishes in the same way.
1.c. Centrifuge the tubes at 1500g for 10 min at 4°C.

1.d. Resuspend the cells in ice-cold PBS and recentrifuged it. Resuspend the cells in TE (pH 8.0) at a concentration of 5×10^7 cells/mL.
1.e. Add 10 mL of lysis buffer per mL of cell suspension. Incubate it for 1 h at 37°C proceed with step 2.

From the suspension of cultured cells:

1.a. Take a centrifuge tube and transfer cells to it. Centrifuge it at 1500g for 10 min at 4°C. Aspirate the supernatant.
1.b. Resuspend the cells in ice-cold PBS in the same volume as that of the original culture and centrifuge it again at 1500g for 10 min at 4°C. Aspirate the supernatant.
Repeat the process once more.
1.c. Suspend the cells in TE at a concentration of 5×10^7 cells/mL.
1.d. Add 10 mL of lysis buffer for each mL of cell suspension and incubate it at 37°C and proceed for step 2.

From tissues:

The tissues contain a lot of fibrous materials, so, they should be powdered to increase the yield of genomic DNA.

1.a. Take 1 g of freshly excised tissue, pour liquid nitrogen to it, and crush it with mortar and pestle.
1.b. Add the powdered tissue to 10 volumes of lysis buffer, mix it well. Transfer it to a 50 mL centrifuge tube and incubate it for 1 h at 37°C and proceed for step 2.

From blood:

1.a. Collect 20 mL of blood (freshly drawn), add a pinch of EDTA to it (in case of frozen blood, allow it to thaw and add PBS to it).
1.b. Transfer it to centrifuge tube and centrifuge at 1300g (3500g for frozen blood) for 15 min at 4°C.
1.c. Aspirate the supernatant, transfer a buffy coat, composed of white blood cells, to a fresh tube. Centrifuge it again. Discard the pellet of red cells.
1.d. Take the buffy coat and resuspend it into 15 mL of lysis buffer.
2. Take another centrifuge tube and transfer the lysate in it, the volume of the tube should be trice of the volume of the content.
3. Add proteinase K to a final concentration of 100 μg/mL. mix it using a glass rod.
4. Incubate it in a water bath preset to 50°C for 3 h with intermittent swirling.
5. Cool the solution and add an equal volume of equilibrated phenol, gently mix it by turning the tube end-over-end in a tube mixture or rolling apparatus, till an emulsion is formed (time required is 10 min to 1 h)

6. Centrifuge it at 5000g for 15 min at RT.
7. Transfer the viscous upper aqueous phase to a fresh centrifuge tube, using a wide bore pipette. Keep in mind that the DNA should be withdrawn very slowly to minimize the hydrodynamic shearing forces.
8. Centrifuge the DNA solution at 5000g for 20 min at RT. Protein and clots of DNA will be at the bottom of the tube. Transfer the supernatant (DNA), to another fresh tube, leaving the proteins and DNA clots.
9. Repeat the process of extraction with phenol twice more and pool the aqueous phase.
10a. If the size of DNA ranges from 150 to 200 kb— transfer the combined aqueous phase to a dialysis bag, with the help of a dialysis tubing dip, close the top of the bag. The volume of the bag should be twice the content of the bag. Perform dialysis at 4°C against 4 l of dialysis buffer. It generally takes 24 h for a process of dialysis to complete. Keep changing the buffer after every 6 h.
10b. If the size of DNA is 100–150 kb—transfer the pooled aqueous phase to a fresh centrifuge tube. Add 0.2 volume of 10-M ammonium acetate. To this, add 2 volumes of ethanol and swirl the tube to mix the contents. This will cause the DNA to precipitate, if it is precipitated in one piece then remove it from the solution using Shepherd's crook. But if, the DNA precipitate gets fragmented then centrifuge the solution at 5000g for 5 min at RT. Wash it twice with 70% ethanol, remove it using an aspirator. Let the DNA air dry.
11. Dissolve the DNA in 1 mL of TE (pH 8.0) per 0.1 mL of cells in step 1.

Analysis and statistics

The obtained DNA is quantitated using a spectrophotometer and the quality of DNA is checked using 0.6% agarose gel.

Precursor techniques

1. 10-M Ammonium acetate: Add 77 g of ammonium acetate to 100 mL of water.
2. Dialysis buffer: 50 mM Tris-Cl (pH 8.0), 10 mM EDTA (pH 8.0). Store at 4°C.
   Stock solutions:
   1 M Tris-Cl (pH 8): Add 12.1 g of Tris base to 80 mL of water and adjust its pH using conc. HCl and make up the volume to 100 mL.
   1 M EDTA (pH 8): Add 29.24 g of EDTA to 80 mL water and adjust its pH to 8 using NaOH pellets and make up the volume to 100 mL.

Take 200 mL of 1 M Tris-Cl (pH 8) and mix it with 40 mL of 1 M EDTA (pH 8) and make up the volume to 4 L using water.
3. Lysis buffer: 10 mM Tris-Cl (pH 8.0), 0.1 M EDTA (pH 8.0), 0.5% SDS, and 20 µg/mL DNase free-pancreatic RNase.
   Add 1 mL of 1 M Tris-Cl (pH 8) and 10 mL of 1 M EDTA (pH 8) to 0.5 g of SDS, mix it and make up the volume to 100 mL. This can be kept under RT.
   Before using add 2 mg of RNase to it. After the addition of RNase, it should be kept under 4°C.
4. TE (pH 8.0): As described earlier.
5. Phosphate buffered saline (PBS): 137 mM NaCl, 2.7 mM KCl, 10 mM Na_2HPO_4, 2 mM KH_2PO_4.
   Dissolve 8 g of NaCl, 0.2 g of KCl, 1.44 g of Na_2HPO_4, and 0.24 g of KH_2PO_4 in 850 mL of water. Adjust the pH to 7.4 using conc. HCl. Autoclave and store it at RT.

Safety considerations and standard

1. Sodium acetate, liquid nitrogen, and conc. HCl should be handled carefully.
2. Be careful while taking out the DNA sample after extracting it with phenol. Take out only the upper aqueous layer. Avoid disturbing interface. If the viscosity of the DNA solution is very high, then it is advised to remove the organic phase using a long pipette.
3. Precool mortar in an ice bucket before adding the liquid nitrogen as suddenly immersing the grinding part of it may rapture it.
4. Care should be exercised while grinding the tissues of human and primate, as aerosols are generated while adding liquid nitrogen to the mortar.
5. Do not desiccate the obtained mammalian DNA sample, as it will be very hard to dissolve it.

Pros and cons

Pros	Cons
Separation and purification of DNA form mammalian cells can be done by this	If not handled carefully DNA molecule will be undergoing hydrodynamic shearing
It is a time taking process. Dialysis requires 24 h	
A large amount of DNA for experiments like southern blotting can be obtained	

Alternative methods

Isolation of DNA using formamide.

Troubleshooting and optimization

Problem	Solution
The DNA is not taken up using the pipettes	Cut the ends of pipettes or use wide bored pipettes
The DNA pellet is not visible	Nothing can be done, start with a new sample, and be careful while extracting DNA from phenol

Summary

1. It is a method to isolate DNA from various cells and tissues.
2. In this protocol, cells are lysed using a lysis buffer and incubated with proteinase K which digest protein and remove the contamination from nucleic acid preparation.
3. This is further subjected to extraction using phenol. The aqueous layer in phenol extraction method contains nucleic acid.
4. Ammonium acetate is used to precipitate the proteins.
5. DNA is precipitated by ethanol.
6. Larger DNA is purified by dialysis.
7. From 5×10^7 cultured cells, the process yields approximately 200 µg of mammalian DNA, 100–150 kb in length. From 20 mL of blood, approximately 250 µg of DNA can be obtained.

Definition

Isolation of DNA from eukaryotic cells.

Rationale

The original method was developed by Richard Palmiter and Ralph Brinsler in 1985. In this method, the tissue/cells are digested with proteinase K to form a homogenous solution of the tissue/cells. Proteinase K digests the proteins. The DNA is extracted by using phenol-chloroform-isoamyl alcohol in the aqueous part of the solution. The DNA in the aqueous part is precipitated using isopropanol and ethanol.

Materials, equipment, and reagents

A. **Reagents**: Ethanol, isopropanol, phenol:chloroform:isoamyl alcohol, phosphate-buffered saline, SNET, TE (pH 8.0), proteinase K.
B. **Glassware/plasticware**: Polypropylene tubes.
C. **Instruments**: Sorvall centrifuge with H1000B and SS-34 rotors (or their equivalents), rocking platform, or shaker incubator.

Protocol

From monolayer of cultured cells:
1.a. Take one batch of cultured dishes with cells grown approximately, more than 80% of confluency.

Place it on an ice bucket. Remove the medium by aspiration and wash the monolayer of cells with ice-cold PBS. Aspirate it and again repeat the washing process quickly.
1.b. Add 1 mL of SNET containing 400 µg/mL proteinase K.
1.c. Transfer the viscous slurry using a rubber policeman to a polypropylene tube.
1.d. Proceed with step 2.
From the suspension of cultured cells:
1.a. Take a centrifuge tube and transfer cells to it. Centrifuge it at 1500g for 10 min at 4°C. Aspirate the supernatant.
1.b. Resuspend the cells in ice-cold PBS in the same volume as that of original culture and centrifuge it again at 1500g for 10 min at 4°C. Aspirate the supernatant.
    Repeat the process once more.
1.c. Suspend the cells in TE at a concentration of 5×10^7 cells/mL.
1.d. Aliquot 200 µL of it in no. of polypropylene tubes.
1.e. Lyse the cells with SNET containing 400 µg/mL proteinase K.
1.f. Transfer it to a polypropylene tube.
1.g. Proceed with step 2.
From mouse tail:
1.a. Cut the tail from 10 days old suckling animal or as per the planned experiment.
1.b. Put it in a polypropylene tube. Add an appropriate amount of SNET containing 400 µg/mL proteinase K.
1.c. Proceed with step 2.
2. Incubate the tubes overnight at 55°C with agitation in a shaking incubator or on a rocking platform so that the contents get mixed properly. After incubation, the color of the contents should be milky-gray, and there should not be any visible tissue left in it.
3. To this add an equal volume of phenol:chloroform:isoamyl alcohol and keep it on a rocking platform for 30 min at RT.
4. Centrifuge the sample in a microfuge tube at max speed for 5 min in a microfuge. Transfer the aqueous phase to a fresh microfuge tube (in case, if the samples are in 17 × 100 mm polypropylene tubes, centrifuge it at 666g for 5 min at RT).
5. Add an equal volume of isopropanol to it. This will precipitate the DNA which can be collected by centrifugation at max speed for 15 min in a microfuge tube (in case, the samples are in 17 × 100 mm polypropylene tubes, centrifuge it at 13,250g for 15 min at RT).

6. Remove the supernatant. Wash the DNA pellet using 70% ethanol by centrifuging it at max speed for 5 min.
7. Remove ethanol and allow the pellets to air dry for 20–25 min at RT. Do not allow the pellet to overdry as it will be difficult to dissolve the DNA pellet.
8. Add 0.5 mL of TE (pH 8.0) and allow it to dissolve it on a rocking platform overnight.
9. Store it at −20°C.

Analysis and statistics
The obtained DNA is quantitated using a spectrophotometer and the quality of DNA is checked using 0.6% agarose gel.

Precursor techniques
1. SNET: 20 mM Tris-Cl (pH 8.0), 5 mM EDTA (pH 8.0), 400 mM NaCl, and 1% (w/v) SDS.
   Stock solutions:
   1 M Tris-Cl (pH 8): Add 12.1 g of Tris base to 80 mL of water and adjust its pH using conc. HCl and make up the volume to 100 mL.
   1 M EDTA (pH 8): Add 29.24 g of EDTA to 80 mL water and adjust its pH to 8 using NaOH pellets and make up the volume to 100 mL.
   10 M NaCl: Add 58.5 g of NaCl to water to make a final solution of 100 mL.
   Mix 2 mL of 1 M Tris-Cl (pH 8) with 500 μL of 1 M EDTA (pH 8), add 4 mL of NaCl to it and make up the volume to 90 mL now add 1 g of SDS to it and make up the final volume to 100 mL. Sterilize it by filtering it through a 0.45 μm nitrocellulose filter.
2. TE (pH 8.0): As described earlier.
3. Phosphate buffered saline (PBS): 137 mM NaCl, 2.7 mM KCl, 10 mM Na_2HPO_4, 2 mM KH_2PO_4.
   Dissolve 8 g NaCl, 0.2 g KCl, 1.44 g Na_2HPO_4, and 0.24 g KH_2PO_4 in 850 mL of water. Adjust the pH to 7.4 using conc. HCl. Autoclave and store it at RT.
4. Phenol:chloroform:isoamyl alcohol: As described earlier.

Safety considerations and standard
1. In case of a newborn or 10 days old mouse, processing should be done in a microfuge tube and 0.5 mL of SNET should be added, while if the mouse is older than this, then the samples should be processed in a 17 × 100 mm polypropylene tube and 4 mL of SNET should be added.
2. The amount of tissue required for a newborn I entire tail, for 10 days old mice is one-third of the entire tail length, for a 3-week-old mouse is 10 mm of the tail.

3. We can store mouse tails or other tissues for a few weeks at −70°C in tightly dosed tubes before adding SNET and proteinase K. But it is advised to proceed without delay to digest the samples with proteinase K. Before phenol:chloroform extraction, the completed digests can then be stored indefinitely at −20°C.
4. Be careful while taking out the DNA sample after extracting it with phenol. Take out only the upper aqueous layer. Avoid disturbing interface. If the viscosity of the DNA solution is very high, then it is advised to remove the organic phase using a long pipette.
5. The pellet obtained after ethanol precipitation is quite difficult to observe so it is advised to arrange the tubes in a centrifuge in a particular way, e.g., all tubes can be kept so that their plastic hinge points outward, the pellet obtained will be at the farthest from the center of rotation and it will be easy to locate it.

Pros and cons

Pros	Cons
Separation and purification of DNA form mammalian cells can be done by this	Time taking process
Relatively easy process and steps and reagents required are less	
Does not require low temperature during the whole process	

Troubleshooting and optimization

Problem	Solution
The DNA is not taken up using the pipettes	Cut the ends of pipettes or use wide bored pipettes
The DNA pellet is not visible	Nothing can be done, start with a new sample, and be careful while extracting DNA from phenol

Summary
1. It is a method to isolate DNA from various cells and especially mouse tissues.
2. In this protocol, cells are lysed using lysis buffer, SNET, and incubated with proteinase K which digests protein and removes the contamination from nucleic acid preparation.
3. This is further subjected to extraction with phenol and the aqueous layer composed of nucleic acid.
4. DNA is precipitated by isopropanol.
5. It does not require a low temperature for processing.

Definition
Separation of bands of DNA by agarose gel electrophoresis.

Rationale

Agarose gel electrophoresis is an easy and efficient method to separate, identify, and purify the DNA molecules. The location of DNA can also be determined with this method by staining with fluorescent dyes, which can detect up to 20 pg of double-stranded DNA by examination of the gel under UV. Agarose gels have relatively lower resolution power than polyacrylamide gels but a greater range of separation. In this process, 50 bp to several megabases of DNA can be resolved in agarose gel (most suited for 50–20,000 bp). Agarose is a linear polymer, it comprises alternate D- and L-galactose joined by $\alpha(1\text{-}3)$ and $\beta(1\text{-}4)$ bonds with anhydro bridge between 3 and 6 positions. It gelatinizes to form a three-dimensional mesh of channels of size ranging from 50 to ≥ 200 nm. The rate of migration of the DNA sample depends on various factors as stated in the previous chapter. One of the factors is the size of the DNA sample.

Materials, equipment, and reagents

A. **Reagents**: Agarose, electrophoresis buffer ($1 \times$ TAE or $0.5 \times$ TBE), ethidium bromide or SYBR gold staining solution, $6 \times$ gel-loading buffer, DNA sample, and DNA ladder.
B. **Glassware/plasticware**: Beaker, spatula, pipettes, tips, gel sealing tape, or gel caster.
C. **Instruments**: Microwave oven or boiling water bath, electrophoresis tank.

Protocol

1. Seal the sides of the gel casting tray or fix the tray in a gel caster.
2. Make agarose solution, as required for the separation of desired DNA fragments, by mixing the appropriate amount of agarose with $1 \times$ TAE or $0.5 \times$ TBE buffer, as per the size of the gel cast the concentration required for DNA separation in various size range is given in the chapter. Some commercially available agarose, as MetaPhor agarose, are high-resolution agarose and can be mixed with the regular agarose to separate DNAs that differ in a few bases only. Then heat it till a clear transparent solution is formed. The buffer should occupy less than half of the volume of the flask used. Undissolved agarose is a small lens like chips in the solution. It should be heated until the complete solution become transparent. If the concentration of agarose is more, a longer heating time will be required to completely dissolve it. Use insulated gloves while handling the solution.
3. Let the agarose solution cool so that the temperature of the beaker is bearable on touching or else transfer the agarose solution to a water bath of 55°C, let it cool. Add ethidium bromide to a final concentration of 0.5 μg/mL, mix it gently, so that no bubbles should form.
4. Pour the solution in the gel casting tray, place an appropriate comb in it to form slots for the samples in the gel. The comb should be placed 0.5–1.0 mm above the plate and the gel should be 3–5 mm thick, air bubbles, if any, should be removed by poking them to the corner of the slab. Let it set and harden. While preparing a gel that contains a low concentration of agarose (<0.5%), a supporting gel, of concentration more than 1%, should first be poured, without any wells. This should be allowed to harden and then, the lower percentage of agarose should be poured on the top of the supporting gel. This will reduce the chances of breaking of the gel while handling. Both supporting and low concentration gels should be made from the same batch of buffer and the same concentration of ethidium bromide. Electrophoresis of gels containing low melting agarose or less than 0.5% agarose should run in a cold room to avoid the breakage of gel.
5. The time for a gel to set is approximately 30–45 min at RT. Pour a small amount of electrophoresis buffer on the gel and carefully remove the comb. Detach the gel from the caster (in case a gel caster is used) or remove the tapes from the sides of the gel slab. Mount the gel in the electrophoresis tank.
6. Fill the electrophoresis tank with the same buffer of the same batch as that of the gel. Even small differences in the ionic strength or pH can create fonts in the gel and can hamper the mobility of DNA fragments. The buffer should cover the gel and the level of the buffer should be at least 1 mm above the upper surface of the gel.
7. Add 0.2 volume of $6 \times$ gel loading buffer to the DNA samples. The amount of DNA required to run depends on the number of fragments of DNA in the sample and their size. In a 0.5 cm of the band, approximately 2 ng of DNA can be detected by staining with ethidium bromide. SYBR gold is a more sensitive dye and can detect up to 20 pg of DNA. More than 500 ng of DNA in a 0.5 cm wide band is considered as overloaded and it might result in trailing and smearing.
8. Load the DNA samples using a disposable micropipette. The DNA ladder or reference DNA is generally loaded at the left or right well of the DNA samples. In case, the DNA needs to be recovered from the gel or in Southern hybridization, different tips for different

samples should be used. For simple experiments, the washing of pipette with buffer is enough.

9. Close the lids of the tank and attach the electrical lids in such a way that the wells should be near the negative electrode (black lead) so that the DNA will migrate toward the positive electrode (red lead). Apply a voltage of 1–5 V/cm (distance between the electrodes). When the leads are attached correctly, electrolysis of buffer will occur, and bubbles will form at the electrodes.

10. Run the gel till the dye has migrated to a suitable distance. Then switch off the apparatus, remove the gel from it, and see it directly in the transilluminator. During electrophoresis, the ethidium bromide migrates toward the cathode, i.e., in the opposite direction to that of DNA; therefore, a prolonged run may cause difficulty in detecting the small fragments. In this case, the gel should be soaked in ethidium bromide (0.5 μg/mL) solution.

Note: SYBR Gold is the trade name of a new ultrasensitive dye with a high affinity for DNA and a large fluorescence enhancement upon binding to the nucleic acid. It can detect less than 20 pg of double-stranded DNA, 100 pg of single-stranded DNA, and 300 pg of RNA. Its maximum excitation is at 300 nm and fluorescence emission occurs at 537 nm. It is supplied as a 10,000 × concentrated dye. While using, dilute it to 1:10,000 and soak the gel in it, after electrophoresis. It should not be added to the gel before electrophoresis, as, in its presence, the gel gets harden and cause its distortion.

Analysis and statistics
The fragments of DNA can be detected in a UV transilluminator.

Precursor techniques
1. Electrophoresis buffer:
   1 × TAE—40 mM Tris acetate and 1 mM EDTA
   Stock solution: **50 × TAE**
   Add 242 g of Tris base in 500 mL of water, add 100 mL of 0.5 M EDTA (pH 8.0) (the method of preparation has been mentioned in the previous chapter), now add 57.1 mL of glacial acetic acid, and make up the volume to 1000 mL.
   Working solution: Take 4 mL of 50 × TAE and make up the final volume to 200 mL.
   0.5 × TBE—45 mM Tris borate and 1 mM EDTA.
   Stock solution: **5 × TBE**
   54 g of Tris base, 27.5 g of boric acid, and 20 mL of 0.5 M EDTA (pH 8.0), now make up the volume to 1 L using water.

Working solution: Dilute 20 mL of buffer to 200 mL of the final volume of the buffer by adding water.

2. 6 × gel-loading buffer:

Stock solution	Recipe
0.25% bromophenol blue 0.25% xylene cyanol FF 40% sucrose in water	Weigh 4 g of sucrose, make the volume to 10 mL by adding water, add 25 μg of bromophenol blue and 25 μg xylene cyanol FF, mix it vigorously and store at 4°C
0.25% bromophenol blue 0.25% xylene cyanol FF 15% Ficoll in water	Weigh 1.5 g of Ficoll, make the volume to 10 mL by adding water, add 25 μg of bromophenol blue and 25 μg xylene cyanol FF, mix it vigorously and store at RT
0.25% bromophenol blue 0.25% xylene cyanol FF 30% glycerol in water	Take 3 mL of glycerol, make the volume to 10 mL by adding water, add 25 μg of bromophenol blue and 25 μg xylene cyanol FF, mix it vigorously and store at 4°C
0.25% bromophenol blue 40% sucrose in water	Weigh 4 g of sucrose, make the volume to 10 mL by adding water, add 25 μg of bromophenol blue, mix it vigorously and store at 4°C

3. Ethidium bromide solution (10 mg/mL): Add 1 g of ethidium bromide to 100 mL of water and stir it on a magnetic stirrer for a few hours to make sure the dye has dissolved. Wrap the solution in aluminum foil and keep it at RT.

Safety considerations and standard
1. Acetic acid, ethidium bromide should be handled carefully.
2. Agarose should be completely dissolved otherwise; the nonhomogeneous gel will hamper the movement of the DNA.
3. Be cautious while handling the beaker with molten agarose, wear heat-protective gloves.
4. Be careful while pouring the molten agarose on the gel slab, there should be no bubbles.
5. The combs should not be too close to the base of the gel slab or else the wells may disrupt while removing the comb.
6. Do not overfill the wells, typically a well (0.5 × 0.5 × 0.15 cm) holds 40 μL of the sample. If the loading DNA is more in volume, then concentrate it by ethanol precipitation.
7. Do not add SYBR Gold to the molten agarose.

Pros and cons

Pros	Cons
Separation and purification of DNA can be done by this	The resolution of agarose gel is lower than that of a polyacrylamide gel. DNA molecule larger than 750 kb cannot be resolved efficiently

Alternative methods

Polyacrylamide gel electrophoresis and pulse-field gel electrophoresis.

Troubleshooting and optimization

Problem	Solution
DNA along with the DNA ladder is not visible in transilluminator	Check whether ethidium bromide is added, soak the gel in its solution (5 μg/mL) for 45 min
Only the DNA ladder is visible, but no other DNA is visible under transilluminator	Check the quality of DNA Assure the connection of the electrophoresis tank is in the proper orientation
The amount of DNA to be loaded is higher than the wells can accommodate	Concentrate the DNA then load it

Summary

1. It is a method to identify, separate, and purify the DNA fragments.
2. Agarose gel is made with buffers and agarose, wells are casted to accommodate the samples.
3. The DNA fragments are separated according to the size, as they move toward the positive electrode (as they bear negative charge) through the gel.
4. Intercalating agents like ethidium bromide intercalate between the strands of DNA and are seen under UV light.
5. 50–20,000 bp DNA is resolved best in the gel.

SOUTHERN TRANSFER AND HYBRIDIZATION

This method was devised by Edwin Southern in 1975. It detects a specific DNA sequence in the DNA sample. This method is used to study the organization of genes in the genome. This technique combines two processes of molecular biology-electrophoresis and hybridization. In this method, a large genomic DNA is digested with restriction enzymes and the resulting fragments are subjected to electrophoresis in an agarose gel which separates them according to their size. Separation of DNA fragments is followed by its in situ denaturation and transfer of gel to a solid support, for example, nylon or nitrocellulose membrane. The membrane with the DNA is then hybridized to a labeled oligonucleotide probe. The binding of the DNA to its probe is detected by a detection system, e.g., autoradiography. After comparing the bands generated by the digestion of genomic DNA with restriction enzymes, it is feasible to place a target DNA in a genome.

Over the years, many advancements as the use of nylon membrane, which has the higher binding capacity and is much more durable than nitrocellulose membranes, usage of advanced autoradiography, more efficient methods for DNA transfer as downward capillary transfer and vacuum blotting, efficient blocking agents, and sensitive phosphor imagers have immensely increased the sensitivity and reproducibility in addition to this, it is now possible to fix the DNA covalently to the membrane which reduces the problems due to leaching of nucleic acids from the membrane at the time of incubation at elevated temperatures.

Before moving to the protocol of Southern blotting, let us have a look at a few aspects of the method in detail:

TRANSFER OF DNA FROM GEL TO SOLID SUPPORT

It is a very important step in the process and usually includes five methods for the transfer:

1. Upward capillary transfer: In this method, the DNA fragments from the gels are carried in an upward flow of liquid to a solid support. The support membrane lies on the upper side of the gel. With the help of capillary action, the buffer rises and gets transferred to the upper part. The capillary action is maintained by the stack of towels of dry absorbent papers. Above the papers, a weight is applied to ensure a tight connection between the papers. The rate of transfer of DNA to the support depends on the size of DNA and the percentage of agarose in the gel. From a 0.7% gel, it takes approximately 1 h, for the fragments that are smaller than 1 kb in size, to get transferred to the support membrane. Larger fragments are transferred at a slower rate and their efficiency of transfer is less. A fragment >15 kb in length will take more than 18 h to get partly transferred. In the case of larger fragments, the efficiency of transfer depends on the no. of molecules that escape the gels before the dehydration of gel. During the elution process, the gel gets modified to a substance of rubber-like texture. This occurs as the fluid is drawn

from the reservoir as well as the gel itself, which changes its texture and it becomes difficult for the DNA molecules to pass through it. To avoid this a partial acid/base hydrolysis of DNA molecules before DNA transfer. In acid-base hydrolysis, the DNA gel is exposed to a weak acid followed by a strong base, the treatment of acid will depurinate it partially and that with a strong base will hydrolyze the phosphodiester backbone at the site of depurination.

This results in the formation of fragments that are less than 1 kb in length and can be readily transferred to the support membrane. Care should be taken so that depurination does not proceed for too long, otherwise, DNA will be cleaved to very small fragments that will be too short to bind the membrane efficiently. Acid/base treatment leading to depurination/hydrolysis should be done only when it is already known that the resultant DNA will be 15 kb in size, as the process, due to increased diffusion, it can cause the DNA bands to appear fuzzy.

2. Downward capillary transfer: In this method, the DNA fragments are carried in a downward flow of liquid to the solid support. The support membrane lies on the lower side of the gel. For the downward capillary transfer, different transfer buffers of varied composition, the arrangement of the reservoir, and wicks can be possible. One of these is 0.4 M NaOH and a set up in which the transfer buffer rises from the reservoir, with the help of wick, to the top of the gel, and then pulled through the gel by the stacks of paper towels that lie under the gel. As compared to the upward capillary transfer, the efficiency of this method is approximately 30% more as the DNA molecules travel more efficiently through the gel that is not under the pressure.

3. Simultaneous transfer to two membranes upper and lower: When the concentration of DNA fragment in the gel is quite high then it can be used to transfer in two membranes simultaneously, one is placed above it and the other is below it. Here, the only source of transfer buffer is the liquid trapped in the gel. The efficiency of this type of transfer is quite low; therefore, it allows signal detection only. It is not recommended for sensitive experiments. But it can be used for Southern analysis of plasmids, bacteriophage, etc.

4. Vacuum transfer: In this method, the nucleic acid is transferred under vacuum. Due to the application of vacuum, the transfer is rapidly and efficiently as compared to the capillary transfer. In the vacuum transfer devices, the gel is placed in contact with the membrane, this membrane lies on a porous

screen over a vacuum chamber. The upper reservoir is filled with a buffer which is used to elute the nucleic acids from the gel, the eluted nucleic acid is then deposited on the membrane. For depurinated and denatured DNA (treated with alkali), the rate of transfer is quite high and it can be transferred within 30 min from a <1% agarose gel. When performing carefully, it can enhance the hybridization signal by 2–3 times. The vacuum should be applied evenly and should not be more than 60 cm of water. An increased vacuum can compress the gel and reduce the efficiency of the transfer. The wells should be handled with care as they can break while preparing the gel for transfer. It is recommended to remove the wells if they get broken.

Membranes

In 1961, Hall and Spiegelman used the nitrocellulose as support for DNA, in powdered form. Later it was used as a sheet. However, nitrocellulose has the following drawbacks:

1. Its capacity to bind DNA is low ($\sim$50–100 µg/cm^2) and it depends on the size of nucleic acids. The nucleic acids <400 bases in length bind to the nitrocellulose inefficiently.
2. The nucleic acids do not interact with the membrane covalently, but by hydrophobic interactions; therefore, during hybridization and washing at high temperature, they can leach slowly from the matrix.
3. The membrane can become brittle during baking at 80°C, under vacuum.
4. Nitrocellulose membrane should be stored carefully, as, if humidity is high, the nitrocellulose membrane can absorb moisture and results in wrinkling when humidity is low, it can dry and get charged with static electricity.

Instead of nitrocellulose membranes, nylon membranes can be used for binding purposes. They bind to the nucleic acids irreversibly and are much more durable. They can even be repaired when damaged. The immobilized nucleic acids can be sequentially hybridized to various probes, without getting damaged.

Two types of nylon membrane: Neutral or unmodified nylon membrane and charge-modified nylon membrane, as it comprises the amine group, is also called positively charged nylon. Both the membranes bind single and double-stranded nucleic acids. Although the charged nylon membrane has a higher capacity to bind nucleic acid, it also gives a higher background because of the nonspecific binding of the phosphate group (negative charged) of RNA and DNA. This problem can be circumvented by using various blocking agents in a high amount, in prehybridization and hybridization steps. The commercially available nylon membranes comprise different extent and type of charge as well as the density of nylon mesh. Therefore, it is advisable to follow the instructions provided by the manufacturer to get the best result. The membranes for Southern blotting can be compared as follows:

Property	Nitrocellulose Membrane	Neutral and Charged Nylon Membrane
Capacity ($\mu g/cm^2$)	80–120	100 (for neutral); 400 (for charged)
Length of nucleic acid for max binding	More than 400 bp	More than 50 bp
Transfer buffer	High ionic strength at neutral pH	Low ionic strength at a wider range of pH
Immobilization	Baking at under vacuum for 2 h	Baking at 70°C for 1 h, no need of vacuum or mild alkali or UV radiation at 254 nm

Let us have a look at protocol for the technique.

Definition
To perform Southern blotting of the genomic DNA.

Rationale
Genomic DNA is first digested with one or more restriction enzymes resulting in DNA fragments of various sizes. These fragments are then subjected to electrophoresis through a standard agarose gel to separate them according to their size. This is followed by in situ denaturations of DNA and its transfer from the gel to a solid support. During the transfer to the membrane, the relative position of DNA bands is preserved. The DNA is then fixed to the membrane and is hybridized with a probe.

Materials, equipment, and reagents
A. **Reagents**: Alkaline transfer buffer, denaturation solution, neutralization buffer I, neutralization buffer II, neutral transfer buffer, $6 \times$ gel loading buffer, agarose, phosphate-SDS washing solution I, phosphate-SDS washing solution II, hybridization buffer, 1-M sodium phosphate (pH 7.2), poly-A RNA (10 mg/mL), probe DNA or RNA, salmon sperm DNA (10 mg/mL).
B. **Glassware/plasticware, others**: Glass baking dishes, glass rod, glass plate, a transparent ruler with fluorescent marking, large-bore pipettes, nylon or nitrocellulose membrane, thick blotting paper (Whatman 3 MM).
C. **Instruments**: Cross-linking chamber or microwave oven or vacuum oven, rocking platform or shaker incubator, hybridization chamber, incubator shaker, and water bath.

Protocol
1. Digest the required amount of genomic DNA with restriction enzymes (one or more). While handling the mammalian DNA use large-bore yellow pipette tips. The points to be remembered during digestion have been mentioned at the end of the protocol.
2. After digestion concentrates the DNA fragments, by ethanol precipitation. Dissolve it in approximately 25 μL of TE (pH 8.0). Residual ethanol may hinder the loading of DNA. Ethanol can be easily removed from the dissolved DNA solution by heating it to 70°C for 10 min. It will also disrupt the base pairing between cohesive ends of fragments resulted from restriction digestion.
3. Measure the amount of DNA. Take an appropriate amount of it in a microfuge tube and add 0.15 volume of $6 \times$ loading buffer to it and load it on a 0.7% agarose gel. The gel may contain ethidium bromide as a regular gel, but the inclusion of ethidium bromide in the gel may distort the DNA, thereby affecting its rate of migration. This will reduce the accuracy in the measurement of DNA fragments. The DNA well should large enough to contain the whole DNA in it. For genomic DNA mostly large gel of size $20 \times$, 20×0.5 cm is used.
4. Sometimes it may be quite difficult to load the DNA in the gel, this problem arises because the DNA is not completely digested and the high-molecular-weight DNA may remain after digestion; therefore digestion should be carried out as described at the start of this protocol. As stated above the DNA should be dispersed homogenously and restriction enzyme should be chosen wisely, for example, the fragments that arise after digestion of genomic DNA with Not I will be very large and it will be problematic to load them. After digestion, if the DNA were stored at 4°C, then heat the DNA to 56°C for 2–3 h before loading. Load the DNA samples slowly and allow the gel to stand for a few minutes so that the DNA is dispersed homogenously in the well.
5. The voltage should be low during electrophoresis (<1 V/cm) to let migrate the DNA very slowly.

6. Stain the gel with ethidium bromide or SYBR Gold after electrophoresis and photograph the gel. Determine the distance traveled by the bands using a ruler with the gel. The DNA can be stored at 4°C after wrapping the gel in saran wrap at this step.

7. Denature the DNA and transfer it using one of the below methods.

8. Transfer the gel in a glass baking dish and trim away the unused portion of DNA including the upper area of the wells. Enough of wells should be attached to the gels so that after the transfer, lanes can be marked on the membrane. Always mark the gel at any side at the bottom and note it down in the notebook to remember the orientation of the gel. The molecular weight markers or DNA ladder may contain some sequence complementary to the probe which can give a signal in autoradiogram and may be puzzling; therefore, it is advised to cut off the area containing ladder or marker.

9. If the desired fragment is larger than 15 kb. It is recommended to depurinate and denature the DNA (as discussed in the previous section of the chapter). Before denaturing the DNA, the gel should be depurinated. Depurination makes the phosphate backbone susceptible to subsequent cleavage. This can be done by soaking in gel in a large amount of 0.2 N HCl until the bromophenol blue and xylene cyanol (of the loading buffer) turns yellow and yellow-green, respectively. As they get changed to the desired color, rinse the gel several times with deionized water. Depuration depends on the diffusion of hydroxyl ions (H^+) therefore, the DNA molecules that are at a different level of the gel cannot be homogenously depurinated. The reaction is difficult to control and it can result in a reduction in the size of DNA and subsequently, the strength of the signal; therefore, depuration is not advisable for fragments lesser than 20 kb.

10. Denaturation is done after depuration by the following methods:
    a. For uncharged membranes:
       • Take denaturation solution in a tray, soak the gel in it for 45 min at RT on a rotatory platform.
       • Rinse the gel in deionized water, then neutralize it by soaking it in neutralization buffer I for 30 min at RT on a rotatory platform. After 30 min, change the neutralization buffer and continue to soak it for 15 more min.
    b. For charged membranes:
       • Soak the gel in alkaline transfer buffer on a rotatory platform for 15 min at RT.

• Change the solution and continue the soaking for 20 more min on a rocking platform.

Preparation of membrane:

11. Cut a piece of nylon or nitrocellulose membrane and thick blotting paper using a fresh scalpel. The size of the membrane should be 1 mm larger than that of the gel in each dimension while the blotting paper should be of the exact size of the gel. While cutting it uses a pair of gloves and blunt-ended forceps as any trace of oil (when the membrane is touched by hands) on the membrane will hinder the wetting of paper.

12. Take a dish and pour deionized water in it, float the membrane in it to get it wet. Take it out and immerse it in the transfer buffer for 5 min. Cut the membrane on the same side as that of gel in step 8. Ensure that the membrane is wet evenly or else change it with a new one and repeat the process of floating. The transfer of DNA in an unevenly wet membrane is not reliable. Uneven wetting is not a problem in the case of the nylon membrane.

    Note: Autoclaving the membrane between pieces of 3-mm paper saturated with $2\times$ SSC will result in an evenly wet membrane.

    Assembly of transfer apparatus:

    To transfer DNA to uncharged membrane neutral transfer buffer ($10\times$ SSC or $10\times$ SSPE) is used.

    To transfer DNA to charged nylon membrane alkaline transfer buffer (0.4 NaOH with 1 M NaCl) is used.

13. Place a piece of thick blotting paper on a glass plate to form support that is longer and wider than gel. Take a baking dish and place the support in it.

14. Pour transfer buffer in an appropriate amount so that the level of liquid reaches the top of the support. Smooth out the air bubble, if any, trapped in below the blotting paper.

15. Take out the gel from the solution in step 10 and place it on the support in an inverted position (the underside of the gel should be on top). There should not be any air bubbles trapped between the gel and blotting paper.

16. Form a barrier with Saran Wrap or Parafilm surrounding the gel so that the liquid can be prevented to flow directly from the reservoir to the paper towels placed on the top of the gel. The paper towels should be stacked precisely or else they will sag near the edge of the gel, may touch the wick, and ultimately result in the inefficient transfer of the DNA.

17. Wet the top of the gel using a transfer buffer and place the wet membrane on the top of the gel. Placing should be done carefully, so that bubbles should not be trapped between the gel and membrane. The membrane can be placed by touching one corner of the membrane to the gel and lowering it slowly.

18. Take two pieces of blotting paper and wet it in the transfer buffer, place them on the top of the membrane. Remove any trapped air bubble between them.

19. Cut numerous paper towels approximately 8 cm in height and place it above the blotting paper. The size of paper towels should be just smaller than that of blotting paper. Put a glass plate above the paper towels and a weight of approximately 400 g above the glass plate. The weight should not be very heavy otherwise the gel would compress. Compression of gel will squeeze the water out of the gel leading to a matrix that will retard the movement of DNA from gel to membrane and the efficiency of transfer will be greatly reduced. The aim of the entire set up is that the buffer from the reservoir is absorbed by the paper towels and blotting papers placed above the gel, and this leads to the transfer of DNA to the membrane which binds to it. Be careful while setting this up and do not disturb it.

20. Allow the transfer to proceed for 8–24 h. paper towels can be replaced on wetting.

21. Remove the blotting paper and paper towels above the gel. Take out the gel with the membrane, place it on a dry sheet of blotting paper, and using a lead pencil or a ballpoint pen, mark the gel slot on the membrane.

22. Remove the membrane from the gel and proceed for the fixation step. Meanwhile, stain the gel with ethidium bromide at 0.5 μg/mL for 45 min and visualize the DNA in a transilluminator.

    DNA fixation to the membrane:

23. The steps of fixation of DNA depends on the type of membrane can be summarized in the following table:

Type of Membrane	Type of Transfer	Type of Fixation
Positively charged nylon	Alkaline transfer	No fixation required
Uncharged and positively charged nylon	Neutral transfer	UV irradiation, baking in a vacuum

24. Soak the membrane in one of the following solutions:

    For neutral transfer: In 6 × SSC for 5 min at RT.

    For alkaline transfer: In neutralization buffer II for 15 min at RT.

    Fixation by baking in vacuum oven:

    • Take out the membrane from 6 × SSC and drain away excess fluid. Place it flat on a paper towel to dry for 30 min at RT.

    • Take a sheet of blotting paper place the membrane on it again put a blotting paper over it. Bake it for 30 min to 2 h at 80°C. Do not overbake it as overbaking can cause the nitrocellulose membrane to become brittle. In the step of neutralization of gel, if it was not completely neutralized then the membrane will turn yellow during baking and can be damaged easily. The nonspecific hybridization will also increase.

    Fixation in a microwave oven: (not recommended in Southern blotting as it attenuates the signal).

    • Take out the damp membrane and put it on a dry blotting paper.

    • Heat the membrane for 2–3 min at full power (750–900 W) in the microwave oven.

25. Cross-linking by UV rays:

    • Take out the damp membrane and put it on a dry blotting paper.

    • Irradiate the membrane at 254 nm. This should be done as per the manufacturer's protocol (for nylon membrane). It is mostly advised to irradiate it to a total of 1.5 J/cm^2 for damp membrane and 0.15 J/cm^2 for dry membrane. This greatly increases the hybridization signal. The process aims to cross-link a small fraction of the thymine of DNA with positively charged amine groups on the surface of the membrane. Over radiation results in covalent attachment of a high proportion of thymine which reduces the hybridization signal.

26. Proceed for the hybridization with a probe.

27. Place the membrane in a tray containing 6 × SSC or 6 × SSPE for 2 min. The membrane should be completely wet.

28. Prehybridization: There are three modes by which it can be prehybridized: For hybridization in heat seal bag, in a roller bottle, and in a plastic container. The description of each is given in the following:

    • For hybridization in a heat-sealable bag: Put the membrane in a heat-sealable bag and add a prehybridization solution at the volume of $0.2\,mL/cm^2$ of the membrane. Squeeze air from the bag.

- Double seal the open end with a heat sealer, and ensure the integrity of seal by gently squeezing the bag. Submerge the bag in a water bath at an appropriate temp. for 1–2 h (42°C for solvents containing 50% formamide; 65°C for phosphate-SDS solvents; and 68°C for aqueous solvents).

- For hybridization in a roller bottle: Roll the membrane and place it inside a hybridization roller bottle along with a mesh of plastic provided by the manufacturer. Add a prehybridization solution at the volume of $0.1\,mL/cm^2$ of the membrane. Close the bottle tightly in a hybridization oven at an appropriate temp. for 1–2 h (42°C for solvents containing 50% formamide; 65°C for phosphate-SDS solvents; and 68°C for aqueous solvents).

- For hybridization in the plastic container: Put the wetted membrane in a plastic container and add a prehybridization solution at the volume of $0.2\,mL/cm^2$ of the membrane.

- Seal the box and place it on a rocking platform in an incubator at an appropriate temp. for 1–2 h (42°C for solvents containing 50% formamide; 65°C for phosphate-SDS solvents; and 68°C for aqueous solvents).

29. In case the radiolabelled probe is double-stranded DNA, it should be denatured by heating at 100°C. Rapidly chill it under ice-cold water. The probe can also be denatured by adding 0.1 volume of 3 N NaOH. Keep it for 5 min at RT, then rapidly chill it under ice-cold water. Add 0.05 volume of 1 M Tris-Cl (pH 7.2) and 0.1 volume of 3 N HCl. This can be stored in ice water until needed.

30. Hybridization: This can be carried out in any of the three methods:
   - For hybridization in a heat-sealable bag: Remove the bag from the water bath, open the bag by cutting it at one corner and pour off the prehybridization buffer.
   - Add to the probe with fresh prehybridization buffer, squeeze the air as much as possible and reseal the bag with as few air bubble bubbles as possible. Reseal it in another plastic bag to avoid radioactive contamination of a water bath. Incubate it in a water bath for an appropriate period.
   - For hybridization in a roller bottle: Pour off the prehybridization buffer and add the probe with

an appropriate amount of prehybridization solution. Incubate it for the appropriate time for hybridization.

- For hybridization in the plastic container: Transfer the membrane to a hybridization bottle and add the probe with an appropriate amount of prehybridization solution. Incubate it for the appropriate time for hybridization.

31. Washing the membrane: For hybridization in heat sealable bag: wear gloves and then remove the bag from the water bath. Remove the outer bag and cut one of the corners of the inner bag. Pour out the hybridization buffer in a suitable container, this hybridization buffer containing probe should be carefully discarded. Further cut the bag and remove the membrane. For bottle one (plastic or roller) carefully take out the membrane with the help of forceps.

32. Submerge the membrane in a tray containing an excess of $2 \times SSC$ and 0.5% SDS at RT on a rocking platform. When hybridization is done with phosphate-SDS solution, place the membrane in excess of phosphate-SDS washing solution I at 65°C on a rocking platform. Repeat this once more. Never allow the membrane to dry out during any stage of the washing process.

33. Pour off the first rinse solution and add an excess of $2 \times SSC$ and 0.1% SDS in the tray. Incubate it on a rocking platform for 15 min at RT. When hybridization is done with phosphate-SDS solution, then rinse the membrane for eight times in excess of phosphate-SDS washing solution 2 at 65°C and skip the step no. 34 and 35 directly move to step no. 36.

34. Pour off the above rinse solution, and add an excess of $0.1 \times SSC$ with 0.1% SDS. Incubate for 30 min at 65°C for 4 h on a rocking platform. During the washing step, monitor the amount of radioactivity periodically. The part of the membrane which does not contain DNA should not emit any signal. If in the process mammalian DNA is used and it is hybridized to a single copy probe, then the whole membrane, including the part containing DNA will not emit any signal detectable by mini-monitor.

35. Wash the membrane with $0.1 \times SSC$ at RT.

36. Remove the excess liquid from the membrane using a paper towel and place the damp membrane on a sheet of Saran Wrap and cover it. Expose the membrane to X-ray film for 16–24 h at −70°C alternatively, expose it to the phosphorimager plate for 1–4 h.

Analysis and statistics

On developing the X-ray film, bands appear at the corresponding position. This indicates the bind of the probe to the target DNA.

Precursor techniques

1. Alkaline transfer buffer: 0.4 N NaOH and 1 M NaCl.

   Stock solution:

   1 N NaOH: Add 40 g of NaOH in water to make a solution of 1 L.

   5 M NaCl: Add 58.5 g of NaCl in water to make a solution of 200 mL.

   Take 200 mL of NaOH solution and add it to 100 mL of NaCl solution and make the volume to 500 mL.

2. Denaturation solution: 1.5 M NaCl and 0.5 M NaOH.

   Take 150 mL of 5 M NaCl solution and 250 mL of 1 N NaOH, make the final volume of 500 mL

3. Neutralization buffer I: 1 M Tris (pH 7.4) and 1.5 M NaCl.

   Stock solution:

   2 M Tris-Cl (pH 7.4): Add 242 g of Tris base to 800 mL of water and adjust its pH using conc. HCl and make up the volume to 1000 mL.

   Take 250 mL of 2 M Tris-Cl (pH 7.4) and 150 mL of 5 M NaCl, make up the volume to 500 mL.

4. Neutralization buffer II: 0.5 M Tris-Cl (pH 7.2) and 1 M NaCl.

   Take 125 mL of 2 M Tris-Cl (pH 7.2) and add 100 mL of 5 M NaCl and make up the volume to 500 mL.

5. Phosphate-SDS washing solution 1: 40-mM sodium phosphate buffer (pH 7.2), 1 mM EDTA (pH 8.0), 5% SDS, and 0.5% fraction-V-grade bovine serum albumin.

6. Phosphate-SDS washing solution 2: 40-mM sodium phosphate buffer (pH 7.2), 1 mM EDTA (pH 8.0), and 5% SDS.

7. Solution for hybridization in aqueous buffer: 6× SSC (or 6× SSPE), 5× Denhardt's reagent, 0.5% SDS, 1 μg/mL poly (A), and 100 μg/salmon sperm DNA. After thorough mixing, filter the solution using a 0.45 μm disposable cellulose-acetate membrane.

   Note: SSPE has EDTA which is a better chelator of divalent ions than citrate. Therefore, it will be more efficient in inhibiting the DNase activity. DNase can decrease the concentration of target DNA and probe.

8. Solution for hybridization in formamide buffer: 6× SSC (or 6× SSPE), 5× Denhardt's reagent, 0.5% SDS, 1 μg/mL poly (A), 100 μg/salmon sperm DNA, and 50% formamide. After thorough mixing, filter the solution using a 0.45 μm disposable cellulose-acetate membrane.

9. Solution for hybridization in formamide buffer: 6× SSC (or 6× SSPE), 5× Denhardt's reagent, 0.5% SDS, 1 μg/mL poly (A), 100 μg/salmon sperm DNA, and 50% formamide. After thorough mixing, filter the solution using a 0.45 μm disposable cellulose-acetate membrane.

10. Solution for hybridization phosphate-SDS buffer: 0.5-M sodium phosphate, 1 mM EDTA (pH 8.0), 7% SDS, and 1% bovine serum albumin.

11. 1 M Sodium phosphate (pH 7.2).

12. Salmon sperm DNA (10 mg/mL): Dissolve salmon sperm DNA (sodium salt of Sigma type III) in the water at a concentration of 10 mg/mL, and it can be dissolved easily by keeping it in a magnetic stirrer for 2–4 h at RT. Extract it with phenol-chloroform. Pass the aqueous phase obtained from the extraction through a 17-gauge hypodermic needle, 12 times rapidly. Add 2 volumes of ice-cold ethanol and precipitate the DNA, recover it by centrifugation, and dissolve at a concentration of 10 mg/mL in water. Check its concentration using a spectrophotometer. Boil the solution for 5–10 min and store at −20°C.

13. 20× SSC: Dissolve 175.3 g of sodium chloride and 88.2 g of sodium citrate in 800 mL of water. Adjust the pH to 7.0 using concentrated HCl. Add water to make up the volume of the solution to 1 l. Sterilize it by autoclaving. The final concentration of sodium chloride is 3 M and that of sodium citrate is 0.3 M.

14. 20× SSPE: Dissolve 175.3 g of sodium chloride, 27.6 of sodium dihydrogen phosphate monohydrate ($NaH_2PO_4 \cdot H_2O$), and 7.4 g of EDTA in 800 mL of water. Adjust the pH to 7.4 using NaOH, sterilize the solution by autoclaving. The final concentration of sodium chloride is 3.0 M that of NaH_2PO_4 is 0.2 and 0.02 M EDTA.

Safety considerations and standard

1. Handle the reagents carefully and take precautions as told in the protocol (in detail) or as per the manufacturer's protocol.

2. In the case of the southern analysis of genomic DNA, each lane comprises approximately 10 μg DNA, the recommended amount of radiolabeled probe is 10–20 ng/mL with ≥10^9 cpm/μg. In case of cloned fragments of DNA, where each lane comprises approximately 10 ng of DNA, the recommended amount of radiolabelled probe is 1–2 ng/mL, with specific activity 10^6–10^9 cpm/μg.

Pros and cons

Pros	Cons
Any type of DNA can be detected by this process	Time taking process

Troubleshooting and optimization

Problem	Solution
Blotchy background over the entire membrane	Prehybridize the membrane for a longer time
	Make sure that the membrane is wet during the whole process
	Do not allow the SDS to precipitate from the solutions containing SDS. This can be avoided by heating the solution to 37°C and then add SDS
	Do not allow the paper towels to get wet completely, replace them with fresh towels during the process
	Increase the concentration of SDS to 1% in all steps and use an uncharged nylon membrane
	While using formamide make sure it is not yellow (impure formamide)
Haloes over the entire membrane	This can be due to the presence of bubbles in hybridization or prehybridization solution, so prewarm the solutions and agitate the membrane wherever mentioned
Background concentrated over the lane of nucleic acids	This may be due to improperly denatured DNA. Reboil the salmon sperm DNA which is used in prehybridization/hybridization buffer. Do not allow it to reanneal
Black spot all over the membrane	This occurs due to the sticking of ^{32}P present as inorganic phosphate or pyrophosphate. Do not use old radiolabeled probes, as hydrolysis can occur in it. In prehybridization/hybridization membrane include 0.5% sodium pyrophosphate

Summary

1. It is a method to identify a target DNA with the help of a probe.

2. In this method, DNA is run in agarose gel to get it separated and then transferred to a suitable membrane with the help of capillary action.

3. The membrane is then fixed and subsequently hybridized with a probe in a hybridization chamber.

4. The signal, due to the binding of the probe to the target DNA fixed to the membrane, is detected by autoradiography.

5. 10 µg of genomic DNA and 10 ng of cloned DNA fragments can be detected by this.

Note: Restriction digestion of genomic DNA:

Before starting the process of digestion, the following points should be considered:

a. The amount of digested DNA fragments should be high enough to generate a signal. In the case of mammalian genomic DNA, when a standard length, i.e., more than 500 bp and a probe of a high specific activity ($>10^9$) are used, then approximately 10 µg of DNA loaded in each slot is sufficient to detect single-copy sequences. The amount of DNA can be proportionately reduced when the DNA contains a high copy no. of the sequence of interest.

b. Suitable restriction enzymes should be used. The median size of the substrate DNA should be at least three times more than the median size of generated fragments.

c. The amount of DNA loaded in each lane should be known. After digestion, the DNA fragments should be concentrated by precipitation with ethanol and measured by fluorometry followed by loading in the gel.

d. DNA should be digested evenly. It is quite hard to digest high-molecular-weight DNA as it may form clumps resulting in variation in its local concentrations. The digests are complete. The steps to ensure the homogenous digestion of DNA are described at the end of this protocol.

e. Always use controls to ensure proper digestion with restriction enzymes, transfer, and hybridization of DNA. A series of high-molecular-weight DNA and control should be loaded simultaneously. A small amount (10^{-5}, 10^{-6}, and 10^{-7} µg) of the plasmid with a sequence complementary to probe can be used as control. The control should be loaded at the extreme left or right side of the gel away from the wells occupied by the test samples of high-molecular-weight DNA.

To ensure proper dispersion of DNA, the following points should be considered while setting up the digestion of genomic DNA:

a. Try to set up the reaction of digestion in a total volume of 45 µL. After the dilution of DNA and adding

of restriction enzyme buffer, but before the addition of restriction enzyme, store the reaction at 4°C for several hours.
b. Stir the DNA solution occasionally, using a sealed glass capillary.

c. First add 5 units/μg of DNA then gently stir the solution for 2–3 min at 4°C, then keep it at 37°C.
d. Add the second aliquot of the enzyme (5 units/μg of DNA) 30 min of digestion.
e. Allow the digestion to take place for 8–12 h.

Southern/Northern blotting assembly

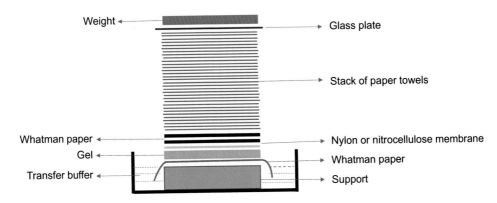

Definition
Isolation of RNA from animal cells.

Rationale
During RNA extraction, the most important step is the inactivation of endogenous RNase. Therefore, many protocols of RNA extraction include the usage of strong denaturants as guanidium salt, which disrupts the cells, solubilizes its components, and denature the endogenous RNase at the same time. The basis of the method is phase separation on the centrifugation of a mixture of aqueous sample and solution of phenol and chloroform. This results in the upper aqueous phase containing nucleic acid and the lower organic phase have protein. The acidic pH provided by sodium acetate causes the DNA to be organic while RNA at the aqueous phase. In neutral condition both RNA and DNA partitioned in the aqueous phase. Guanidinium cyanate aids in the denaturation of proteins (including RNase) and β-mercaptoethanol, and sarcosine too acts as denaturing agents. Then isopropanol (added to the aqueous phase) causes the RNA to precipitate out.

Materials, equipment, and reagents
1. **Reagents**: Liquid nitrogen, chloroform:isoamyl alcohol (49:1, v/v), ethanol, isopropanol, phenol, phosphate-buffered saline (PBS), 2-M sodium acetate (pH 4.0), denaturing solution.

2. **Glassware**: Polypropylene snap cap tubes, pipettes, cuvettes, mortar, and pestle (washed with DEPC-treated water).
3. **Instruments**:   Spectrophotometer,   polytron homogenizer.

Protocol
1. Prepare the sample using one of the following methods:
    For tissues:
    a. Isolate the tissues by dissecting them and placing them in the liquid nitrogen (this snap-frozen tissue can be stored at −70°C. Snap freezing is recommended for the tissues which are rich in degradative enzymes, such as pancreas or gut. In tissues that are not as rich in the enzyme RNase, they can be just minced and transferred to polypropylene tubes).
    b. Transfer approximately 100 mg of frozen tissue to a prechilled mortar containing liquid nitrogen, grind the tissue with the help of a pestle.
    c. Take 3 mL of solution D in a polypropylene snap cap tube, transfer the powdered tissue to it. Homogenize it for 30 s at RT using polytron homogenizer.
    For mammalian cells grown in monolayer:
    a. Aspirate the medium, rinse cells with 10 mL of sterile ice-cold PBS.

b. Aspirate the PBS and add 2 mL of solution D per 90 mm dish.

c. Transfer the lysate to a polypropylene snap cap tube.

d. Homogenize it for 30 s at RT using polytron homogenizer.

For mammalian cells grown in suspension:

a. Centrifuge and harvest the cells at 200g for 5–10 min at RT.

b. Aspirate the medium and resuspend the cells in 2 mL of ice-cold sterile PBS.

c. Again, centrifuge at 200g for 5–10 min at RT. Add 2 mL of solution D per 10^6 cells.

d. Homogenize it for 30 s at RT using polytron homogenizer.

2. Take a fresh polypropylene tube and transfer the homogenate into it. Now add 0.1 mL of 2-M sodium acetate (pH 4.0), 1 mL phenol, and 200 μL of chloroform-isoamyl alcohol per mL of solution D. Cap the tube and mix the contents of the tube by inversion, after every addition.

3. Vortex for 10 s and then incubate the tube for 15 min in ice.

4. Centrifuge it at 10,000g for 20 min at 4°C, transfer the upper aqueous layer with extracted RNA to a fresh tube.

5. Add an equal amount of isopropanol to the tube, mix it well and keep it at −20°C for more than 1 h, so that the RNA gets precipitated.

6. Centrifuge it at 10,000g for 30 min at 4°C.

7. Carefully remove the supernatant and dissolve RNA pellet in 0.3 mL of solution D for every mL of it used in step 1. The pellet is not visible in many cases, or it is loosely attached and can be easily lost. Therefore, the supernatant should be kept in a separate fresh tube and should not be discarded, till the pellet has been checked.

8. Transfer the above solution in a fresh test tube and add 1 volume of isopropanol to it. Keep it at −20°C for more than 1 h.

9. Centrifuge it at max speed in a microfuge for 10 min at 4°C. Wash the pellet twice with 75% ethanol, centrifuge it at max speed for 10 min at 4°C.

10. Decant the supernatant carefully, and let the ethanol evaporate.

11. Add 50–100 μL of DEPC-treated water to the RNA pellet. The addition of SDS to 0.5% of concentration may help in dissolving the pellet. It can also be stored in TE (pH 7.8) with 0.1%–0.5% SDS or DEPC-treated water with 0.1 mM EDTA (pH 7.5).

Note: Storage of RNA:

A. It is preferred to dissolve purified RNA pellet in 50–100 μL of stabilized formamide and store the solution at −20°C. RNA can be recovered from formamide by precipitation with ethanol (4 volumes).

B. SDS (used for dissolving RNA) can hinder many reactions in which RNA is used as a templet, so, it should be removed from RNA by chloroform extraction and ethanol precipitation before any enzymatic treatment.

C. Final RNA preparation may be treated with RNase free DNase, mostly in the case of transfected cells where significant DNA contamination of RNA is detected. In addition to this DNA fragments can be removed by oligo d(T) chromatography.

Analysis and statistics

The quality of RNA can be checked by agarose gel electrophoresis and quantity by spectrophotometer at 260 nm.

Precursor techniques

All the reagents are made in DEPC-treated water.

1. Denaturing solution: 4-M guanidium thiocyanate, 25-mM sodium citrate, 0.5% sodium lauryl sarcosinate, and 0.1-M β-mercaptoethanol.

Stock:

1-M Sodium citrate: 13 g of sodium citrate to a final volume of 50 mL with water.

5% Sodium lauryl sarcosinate: Dissolve 5 g of it to make 100 mL of the final solution.

Dissolve 94.5 g of guanidium thiocyanate to 150 mL of water, add 5 mL of 1-M sodium citrate solution, 20 mL of 5% sodium lauryl sarcosinate, and make up the volume to 200 mL. This can be stored in a dark bottle. Remember that guanidium salts can be precipitated at low temperatures. Before using, add 0.36 mL of β-mercaptoethanol per 50 mL of working solution.

Amount of denaturing solution required is

For 90 mm dish—2 mL.

For 60 mm dish—1 mL.

For 100 mg tissue—3 mL.

2. Chloroform:isoamyl alcohol (49:1): Mix 49 mL of chloroform and 1 mL isoamyl alcohol.

3. 2-M sodium acetate (pH 4.0): Weigh 8.2 g of sodium acetate to 40 mL water, adjust its pH to 4, with the help of acetic acid.

Safety considerations and standards

1. Denaturing solution is very caustic. Wear gloves while handling.

2. Pellets are easily lost. Decant the supernatant into a fresh tube. Do not discard it until the pellet has been checked.

3. To minimize the DNA contamination avoid taking the lowest part of the aqueous phase after extraction with chloroform:isoamyl alcohol.

Alternative methods
By spin columns.

Pros and cons

Pros	Cons
An easy and quick method for extraction of RNA	Not suitable for isolation of RNA from adipose tissue Proteoglycan and polysaccharides may remain with RNA

Summary
1. In this method, the RNA is extracted in aqueous solution with the help of phenol and chloroform.
2. The acidic pH makes the RNA to be in the aqueous phase.
3. Guanidinium cyanate act as a strong denaturant that disrupts the cell and also denatures the proteins.
4. The yield of RNA mainly depends on its source, in tissues it yields 4–7 μg of RNA can be obtained from 1 mg of tissue, while it is 5–10 μg/10^6 cells. A_{260}/A_{280} of the extracted RNA is 1.8–2.0.

NORTHERN BLOTTING
Northern hybridization is a technique that is used to measure the size, abundance, and amount of RNAs. The process was first described by Alwine et al. in 1977, since then it has been modified many times, but the basic protocol remains the same:
1. The isolation of intact mRNA.
2. Its separation in a denaturing agarose gel.
3. Transfer followed by the fixation of RNA to support.
4. Hybridization of immobilized RNA using a probe (complementary to the target sequence).
5. Processing of the membrane to remove the nonspecific binding.
6. Capturing and analysis of the image.

Let us first have a look at the procedure in detail:

Separation of RNA according to their size—The RNA is subjected to electrophoresis through denaturing gel, which will separate it according to their size. Earlier, methyl mercuric hydroxide was used as a denaturing agent, but it is highly toxic and volatile. So, now it has been substituted with glyoxal/formamide or formaldehyde and dimethyl sulfoxide (DMSO). The two methods used to separate RNAs are:

- Electrophoresis of RNA that has been denatured with glyoxal/formamide, through an agarose gel.
- Pretreatment of RNA with formaldehyde and dimethylsulfoxide and electrophoresis of the RNA through the gel with 2.2 M formaldehyde.

However, glyoxal and formaldehyde are toxic. Guanidinium thiocyanate may have an advantage over formaldehyde and glyoxal but is not used in many labs.

The choice of system of electrophoresis depends on the preference of lab.

The method of separation in agarose gel is adapted from McMaster and Carmichael (1977). Glyoxal is also known as ethanedial and diformyl. It is used to eliminate the secondary structure in a single-stranded RNA. It has two aldehyde groups that prevent the formation of intrastrand Watson and Crick bonds by reacting to the amino group of guanosines and form a cyclic compound. There is no need for incorporation of glyoxal in the agarose gel as once the cyclic compound is formed, it remains stable at RT at pH ≤7.0. Therefore, in the absence of secondary structure, the glyoxylate RNA migrates in the gel according to their size.

The electrophoresis in the agarose gel must be carried out at low ionic strength to prevent the renaturation of RNA. Earlier, the buffers used for this purpose was 10 mM phosphate or 40-mM morpholinopropane sulfonic acid (MOPS). These buffers have limited buffering capacity. During the running of agarose gel, an upward shift in pH of electrophoresis buffer occurs at the cathode chamber. If this is allowed to continue, then a steep gradient of pH would form as from the cathode small ions will start migrating along with the gel and these buffers with low buffering capacity will be problematic. The flow of ions results in the removal of glyoxal adduct from RNA. Due to this reason, it was a practice to mechanically recirculate the buffer or to replace it periodically during electrophoresis. This problem can be avoided by using a more stable buffer containing a weak acid and a weak base with similar pK values. In the electrophoresis of RNA staining, the glyoxalated RNA is also a matter of concern. After electrophoresis, staining the gel with ethidium bromide is insensitive due to a high background of nonspecific fluorescence. Staining the RNA during glyoxylation before loading to the gel with ethidium bromide is also not possible as the ethidium bromide reacts with glyoxal. Staining RNA with acridine orange gives a stronger signal than ethidium bromide but extensive washing of the gel is needed. In 1994, Grundemann and Koepsell reported that RNA can be stained effectively, during denaturation with glyoxal, however, the efficiency of hybridization is decreased.

Transfer of RNA and its fixing to solid support—
Two types of transfers are possible which are described as below, the method of fixing too depends on the choice of transfer.

Transfer to a positively charged nylon membrane at alkaline pH:

In alkaline solution, nylon membranes can retain nucleic acids. From agarose gel, RNA can be efficiently transferred in 8.0 mM NaOH with 3 M NaCl. Under this condition, the RNA gets partially hydrolyzed, and its efficiency and speed increase, particularly for large fragments >2 kb in size. Under alkaline conditions, there is no need to fix the RNA by baking or UV irradiation as it gets covalently attached to the membrane and does not require any further fixation. The alkaline transfer has a few drawbacks, for example, background hybridization is quite high, especially when the RNA probe is used. This background further increases, if, the exposure time of nylon membrane to the alkaline solution is more than 6 h. The background can be decreased by a decrease in the duration of transfer and by using an increased amount of blocking agents in prehybridization and hybridization step. It is recommended to use nylon 6 membranes instead of nylon 66 membranes as variability in the hybridization signal has been reported in nylon 66 membrane, after alkaline transfer.

Transfer to uncharged nylon membrane at neutral pH:

In this method, 10× or 20× SSC is used and the RNA is linked to the membrane by baking it or by UV irradiation as stated above.

Materials, equipment, and reagents

A. **Reagents**: 10× BPTE electrophoresis buffer, DMSO, glyoxal, glyoxal reaction mixture, ethidium bromide (200 μg/mL), formaldehyde, formamide, formaldehyde gel loading buffer, MOPS electrophoresis buffer. 10× formaldehyde gel loading buffer: 50% glycerol, 10 mM EDTA (pH 8.0), 0.25% xylene cyanol, 0.25% FF bromophenol blue, 0.1-M ammonium acetate with 0.5 μg/mL ethidium bromide, 0.02% Methylene blue solution in 0.3-M ammonium acetate (pH 5.5), soaking solution for charged membrane and uncharged membrane, transfer buffer for alkaline and neutral transfer, Prehybridization solution, SSC (0.5×, 1×, and 2×) with 1% SDS, and SSC (0.1× and 0.5×) with 0.1% SDS.

B. **Glassware/plasticware, others**: Glass baking dishes, glass rod, glass plate, the transparent ruler with fluorescent marking, large-bore pipettes, nylon or nitrocellulose membrane, thick blotting paper (Whatman 3 MM), and scalpel blade.

C. **Instruments**: Cross-linking chamber or microwave oven or vacuum oven, rocking platform or shaker incubator, hybridization chamber, incubator shaker, and water bath.

Protocol

1. Cast 1.5% agarose gel with 1× BPTE electrophoresis buffer. The combs used in this should have at least four more teeth than the total number of RNA samples. An extra lane is used for loading the RNA ladder and running dyes. For 0.5–8.0 kb RNA, agarose gel of 1.5% is suitable, while, if the RNA is larger than this range use a gel of 1.0%–1.2%.

2. Set up a denaturation reaction comprising glyoxal and RNA. For this in a sterile microfuge tube, mix the following:
   RNA—10 μg. (1–2 μL).
   Glyoxal reaction mix—10 μL.
   Close the tubes and incubate it at 55°C for 60 min. Chill the samples in ice water and give a short spin to bring down all the liquid, for proper denaturation.

3. Simultaneously, put the agarose gel in the electrophoresis apparatus and add 1× BPTE buffer so that the gel gets covered to a depth of nearly 1 mm.

4. Add 1–2 μL of RNA gel loading buffer to the glyoxalated RNA and immediately load it to the wells of the gel. On each side, the last two lanes of the gel should be left and used to load RNA markers/ladders.

5. Start the process of electrophoresis at 5 V/cm till the dye bromophenol blue migrates to approximately 8 cm. Using higher voltage during electrophoresis causes smearing of the samples.

6. Place the gel on Saran Wrap and visualize the RNA using a transilluminator. Align a transparent ruler with a fluorescent marker alongside the gel, so that the photograph can be used to determine the distance traveled by the RNA from their respective wells.

7. Immobilize the RNA on a solid support.

8. The denaturation of RNA can also be done by using formamide and then separated through agarose gels containing 2.2 M formaldehyde. Formaldehyde reacts with the imino group of guanine residue to form an unstable Schiff base. This prevents the formation of intrastrand Watson-Crick base pairing between the bases and thus there is no formation of the secondary structure of RNA. As the Schiff base is unstable so it is necessary to incorporate formaldehyde in either buffer or gel. Agarose gels

comprising formaldehyde less elastic and frangible than normal agarose gel, so it should be handled carefully during shifting from one place to another. Formaldehyde gets dissociated easily from RNA, as compared to the glyoxal; however, bands of RNA in the formaldehyde denaturation are quite fuzzy. The gel comprises 6% or 2.2 M formaldehyde, such a high concentration compensates for the loss of formaldehyde by diffusion during electrophoresis. Alternatively, it can also be avoided by running the gel at higher voltage, i.e., 7–10 V/cm, for a short period. The concentration of formaldehyde gets reduced to 1.1% or 0.66 M at the end of electrophoresis. Ethidium bromide should not be incorporated into the gel containing formaldehyde as the gel is irradiated by UV, because of pinkish glow all over the gel small RNAs could not be detected. Instead, the RNA sample should be heated with a small amount of ethidium bromide before loading it to the gel. If the concentration of ethidium bromide does not exceed 50 μg/mL, there will be no reduction in the efficiency of hybridization.

For running RNA agarose gel with formaldehyde:

1. Prepare 1.5% agarose gel with formaldehyde for 0.5–8.0 kb RNA and 1%–1.2% for larger RNA. For this, take 1.5 g of agarose and add it 72 mL water. Dissolve the agarose by boiling it. Cool the solution to 55°C and then add 10 mL of 10 × MOPS buffer and 18 mL deionized formaldehyde to it. Cast the formaldehyde/agarose gel in a chemical fume hood. The gel should have four more wells than the no. of RNA sample, and an extra lane is used for RNA ladder and running dyes. Allow it to sit for 1 h and then cover it using a Saran Wrap until the samples are ready.

RNA samples should consist of approximately 20 μg RNA in <2 μL of the sample. If RNA has been stored, precipitate it using ethanol and dissolve it in an appropriate volume of DEPC-treated water. Loading more than 20 μg RNA or the presence of salts or SDS in the buffers can cause smearing of RNA during electrophoresis. During the electrophoresis, the pH of buffer changes; therefore, circulate the buffer from one chamber to another either manually or using a pump.

1. Set up the denaturation reaction by adding the following in a sterile test tube:
   RNA (20 μg) 2 μL.
   10 × MOPS buffer 2 μL.
   Formaldehyde 4 μL.

Formamide 10 μL.
Ethidium bromide 1 μL.

RNA size markers should also be treated like the samples. Approx. 20 μg of RNA can be detected in each lane of the gel. Abundant mRNA can be detected by analysis of 10 μg of total cellular RNA, while, in the case of rare RNAs, at least 1.0 μg of poly(A)$^+$ RNA should be loaded in each lane.

2. Mix the above solution and incubate it at 55°C for 1 h. (Some prefer to incubate it for 85°C for 10 min.)
3. Cool it down in ice water for 10 min. Give a short spin so that the liquid comes down to the bottom of the tube.
4. To each sample add 2 μL of 10 × formaldehyde gel loading buffer and again incubate in the ice bucket.
5. Install agarose/formaldehyde gel in the electrophoresis tank (always reserve an electrophoresis apparatus for RNA analysis. Clean the combs, gel casting slab and electrophoresis tank with detergent, wipe with ethanol, and fill a solution of 3% H_2O_2 in it, let it be like that for 10 min. Afterward, rinse everything with DEPC-treated water. During electrophoresis, the pH of the buffer gets changed; therefore, it is necessary to circulate the buffer from one chamber to another manually or using a pump).
6. Add a sufficient amount of 1 × MOPS electrophoresis buffer to it, and the level of the buffer should be 1 mm above the gel.
7. Before loading the samples, run the gel for 5 min at 5 V/cm. After this load the samples, leaving two lanes at either side of the gel. Load RNA markers/size standards to it.
8. Run the gel at 4–5 V/cm, till the bromophenol blue has migrated ~8 cm, this will take approx. 4–5 h.
9. Place the gel on Saran Wrap on a UV transilluminator and align the transparent ruler with fluorescent markers with it. Plot a graph of $\log_{10}$ of the size of RNA fragments in the Y-axis and distance traveled by them in the X-axis, to calculate the sizes of RNA that has been detected by hybridization.
   Preparation of gel for transfer:
   For the transfer to uncharged nylon membrane:
10. Rinse the gel with DEPC-treated water. Then, soak the gel in excess of 0.05 N NaOH for 20 min.
11. Transfer the gel to an excess of 20 × SSC for 40 min. Immediately, the process of step 14.
    For the transfer to charged membrane:
12. Rinse the gel in DEPC-treated water.
13. Soak the gel in 0.01 N NaOH/3 M NaCl, for 20 min. Immediately, the process of step 14.

14. Using a sharp scalpel, cut the unused area of gel. Cut off either of the sides at the corner to mark the orientation of the gel (it is advised to make a practice to cut the same corner so that it will be easy to remember).

15. Place a thick bolting paper on a glass plate to form support that is larger than the trimmed gel both in length and breadth. Drape the paper over the edges of the plate.

16. Place the support in the baking dish. Elevate the support using a neoprene stopper. Fill the dish with suitable transfer buffer (for positively charged membranes 0.01 N NaOH/3 M NaCl, and uncharged membranes $20 \times$ SSC) till the level of buffer reaches the top of the support. Smooth out the bubbles, if any, with a glass rod.

17. With a fresh scalpel cut the piece of nylon membrane approx. 1 mm larger than gel in both dimensions. While cutting it uses a pair of gloves and a pair of blunt-ended forceps as any trace of oil (when the membrane is touched by hands) on the membrane will hinder the wetting of paper.

18. Take a dish of deionized water and float the nylon membrane till it gets wet, then immerse it in $10 \times$ SSC for 5 min. Cut a corner (as done for the gel) to mark the orientation of the membrane. Make sure the membrane is wet after a few min of floating, if it is not evenly wet then change the membrane, as the transfer in an unevenly wet membrane is not reliable.

19. Carefully invert the gel and place it in the center of the wet blotting paper. There should not be any air bubble between the blotting paper and the gel.

20. To prevent the liquid from flowing directly from the reservoir to the paper towels, make sure to surround the gel with Saran Wrap.

21. Wet the top of the gel using a transfer buffer and place the wet nylon membrane on the top of the gel. The cut corner of both the gel and membrane should align. Be careful, do not move the membrane once placed on the gel. There should not be any air bubbles between the gel and membrane.

22. Cut two pieces of thick blotting paper of the same size as that of the gel, wet it in the transfer buffer, and place it on the top of the nylon membrane.

23. Cut numerous paper towels approximately 8 cm in height and place it above the blotting paper. The size of paper towels should be just smaller than that of blotting paper. Put a glass plate above the paper towels and a weight of approximately 400 g above the glass plate. The weight should not be very heavy

otherwise the gel would compress. Compression of gel will squeeze the water out of the gel leading to a matrix that will retard the movement of DNA from gel to membrane and the efficiency of transfer will be greatly reduced. The aim of the entire set up is that the buffer from the reservoir is absorbed by the paper towels and blotting papers placed above the gel. This leads to the transfer of RNA to the membrane which binds to it. Be careful while setting this up and do not disturb it.

24. In the case of the neutral buffer, allow the transfer to take place for not more than 4 h, while in case of alkaline buffer allow it to take place for approximately 1 h.

25. After transfer, dismantle the entire set up, using a ballpoint pen, mark the position of the slots on the membrane. Transfer the membrane to a tray containing an excess of $6 \times$ SSC at 23°C. Put the tray on a rocking platform for agitation (very slow) for 5 min. Meanwhile, the gel can be stained with ethidium bromide (0.5 µg/mL in 0.1 M sodium acetate) for 45 min, to determine the extent of transfer of nucleic acid to the membrane.

26. Take out the membrane from $6 \times$ SSC and let the excess fluid drain away. Place the membrane, with RNA side upwards, on a blotting paper for 1 min.
Staining and fixing:

27. The steps of staining and fixation of RNA depends on the type of transfer. The order of the steps is as follows:
    - For positively charged membranes, the transfer can be alkaline or neutral.
    - For uncharged or positively charge membrane the transfer will be neutral only.

Method of Transfer	Mode of Fixation	Order of Steps for Staining and Fixation
Alkaline	No fixation required	1. Stain membrane with methylene blue 2. Proceed with prehybridization
Neutral	UV 1. Fix RNA with UV 2. Stain	irradiation membrane with methylene blue 3. Proceed with prehybridization
Baking	1. Stain	membrane with methylene blue 2. Fix it by baking 3. Proceed with prehybridization

Staining of the membrane:

Take a glass tray, fill it with methylene blue solution, and transfer the membrane (damp) to the tray. Keep it for 3–5 min so that the membrane gets stained. Take its photograph and then proceed with destaining. Wash the membrane with $0.2 \times$ SSC with 1% SDS for 15 min at RT.

Fixing by baking:

Let the membrane dry in the air and then keep it between two pieces of blotting and bake for 2 h under vacuum at 80°C in a vacuum oven or heat it at full power in a microwave oven of 750–900 W for 2–3 min.

Cross-linking by UV irradiation:

Take a dry blotting paper and keep the damp, unstained membrane on it, and irradiate at 254 nm for 105 s at 1.5T/cm^2. Stain the membrane after irradiation.

Hybridization with the probe:

28. Proceed with the prehybridization step. If the membrane is not used immediately, the membrane can also be stored after thoroughly drying and wrapping it in aluminum foil or blotting paper under vacuum at RT.

29. Incubate the membrane for 2 h at 68°C in 10–20 mL of prehybridization buffer. Incubation can be done in either a commercial rotating wheel or a plastic tiffin box or plastic sealable bag. The only advantage of commercial wheels is that it does not leak.

30. No denaturation is required for single-stranded probes. If a double-stranded probe is being used, and then, denature it by heating the ^{32}P labeled double-stranded DNA, for 5 min at 100°C. Cool it down immediately in ice. It can also be denatured by adding 0.1 M NaOH followed by incubation in ice water of the same (after 5 min) and then adding 0.05 volume of 1 M Tris-Cl and 0.1 volume of 3 N HCl.

31. Add the denatured probe to a direct prehybridization solution. Incubate for 12–16 h at RT. To detect low-abundance mRNA, use at least 0.1 μg with specific activity more than 2×10^8 cpm/μg. (When the probe is not homologous to the target gene then low stringency hybridization is carried out at low temperature (37–42°C) in a hybridization buffer comprising 50% formamide, 0.25 M sodium phosphate (pH 7.2), 0.25 M NaCl, and 7% SDS.)

32. Take a plastic box and fill it with 100–200 mL of $1 \times$ SSC and 0.1% SDS. Transfer the membrane from the plastic bag to the plastic box. Put it on a shaker and agitate it gently for 10 min. Be careful, do not let the membrane dry out at any stage of washing. If a single-stranded probe is used, increase the concentration of SDS to 1% washing buffer.

33. Take another plastic box with $0.5 \times$ SSC with 0.1% SDS prewarmed at 68°C and agitate it at 68°C for 10 min. Repeat it two more times. (Following low stringency hybridization in buffers containing formamide, rinse the membrane at 23°C in $2 \times$ SSC. This should be followed by successive washing in $2 \times$ SSC, $0.5 \times$ SSC with 0.1% SDS, and $0.1 \times$ SSC with 0.1% SDS for 15 min at 23°C each. Finally, $0.1 \times$ SSC with 1% SDS at 50°C.)

34. Take out the membrane and put it on a blotting paper, expose it to X-ray film for 24–48 h at $-20°$ C with tungstate-based screens to intensify the signals, or scan it using a phosphorimager.

The membranes can be reused for hybridization with other probes after stripping the nylon membrane by treating it with 10 mM Tris-Cl (pH 7.4), 0.2% SDS, preheated to 70–75°C or 50% formamide, $0.1 \times$ SSC, and 0.1% SDS preheated to 68°C.

Precursor techniques

All the reagents are made in DEPC-treated water.

1. $10 \times$ MOPS electrophoresis buffer: 0.2 M MOPS (pH 7.0), 20-mM sodium acetate, 10 mM EDTA (pH 8.0) **Stock:**

   1-M sodium acetate—Dissolve 4.1 g of sodium acetate to make a final volume of 50 mL of solution with DEPC-treated water.

   1 M EDTA (pH 8): add 14.6 g of EDTA to 40 mL water and adjust its pH to 8 using NaOH pellets and make up the volume to 50 mL.

   Dissolve 41.8 g of MOPS (3[*N*-morpholino] propane sulfonic acid) in 750 mL of water, adjust its pH to 7.0 using NaOH, add 20 mL of 1 M sodium acetate to it, and 10 mL of 1 M EDTA to it. Add DEPC-treated water to make a final volume up to 1 L. Filter it through 0.45 μm filter and keep it at RT protected from light. The buffer turns yellow with age and on exposure to light or after autoclaved. Straw-colored buffer is OK but it is not recommended to use a darker colored buffer.

2. $10 \times$ BPTE buffer: 100 mM PIPES (piperazine-1,4bis [2-ethanesulphonic acid]), 300 mM Bis-Tris, 10 mM EDTA (pH 8.0).

   Weigh 3 g of PIPES (free acid) and add 6 g of Bis-Tris (free base) and 2 mL of 0.5 M EDTA (pH 8.0) to 90 mL of distilled water, then add DEPC to the solution with final concentration 0.1%, for 1 h at 37°C, and then autoclave.

3. Soaking solution:

   For charged membrane—0.01 N NaOH and 3 M NaCl

   Stock:

   1 M NaOH—Add 4 g of NaOH to make 100 mL of the solution to water.

   Take 17.55 g of NaCl, add 1 mL of 1 M NaOH to it and make the volume to 100 mL.

   For uncharged membrane—0.05 N NaOH—dilute 5 mL of 1 M NaOH to 100 mL of solution.

4. Transfer buffer:

   For alkaline transfer—0.01 N NaOH and 3 M NaCl—same as mentioned above.

   For neutral transfer—20× SSC: Dissolve 175.3 g of sodium chloride and 88.2 g of sodium citrate in 800 mL of water. Adjust the pH to 7.0 using concentrated HCl. Add water to make up the volume of the solution to 1 L. Sterilize it by autoclaving. The final concentration of sodium chloride is 3 M and that of sodium citrate is 0.3 M.

5. Prehybridization solution: 0.5-M sodium phosphate buffer (pH 7.2), 7% SDS and 1 mM EDTA (pH 7.0).

   Stock:

   1-M Disodium phosphate (Na2HPO4): Weigh 14.2 g of disodium phosphate and add it to water to make a solution of 100 mL.

   1-M Sodium dihydrogen phosphate (NaH2PO4): Weigh 12 g of sodium dihydrogen phosphate and add it to water to make a solution of 100 mL.

   Add 34.2 mL of 1-M disodium phosphate and 15.8 mL of sodium dihydrogen phosphate EDTA to make a solution of 1 L with water.

6. SSC (0.2×, 0.5×, 1×, and 2×) with 1% SDS—add 5, 12.5, 25, and 50 mL of 20× SSC, respectively, add 5 g of SDS and make up the solution to 500 mL.

7. SSC (0.1× and 0.5×) with 0.1% SDS—add 2.5 mL and 12.5 of 20× SSC, respectively, and add 0.5 g of SDS to make up the solution to 500 mL.

Safety considerations and standard

1. Handle all the reagents carefully especially formaldehyde which is teratogenic.
2. Take precautions as told in the protocol (in detail) or as per the manufacturer's protocol.

Pros and cons

Pros	Cons
Any type of RNA can be detected by this process	Time taking process

Troubleshooting and optimization

Problem	Solution
Blotchy background over the entire membrane	Prehybridize the membrane for a longer time
	Make sure that the membrane is wet during the whole process
	Do not allow the SDS to precipitate from the solutions containing SDS. This can be avoided by heating the solution to 37°C and then add SDS
	Do not allow the paper towels to get wet completely, replace them with fresh towels during the process
	Increase the concentration of SDS to 1% in all steps and use an uncharged nylon membrane
	While using formamide make sure it is not yellow (impure formamide)
Haloes over the entire membrane	This can be due to the presence of bubbles in hybridization or prehybridization solution, so, prewarm the solutions and agitate the membrane wherever mentioned
Background concentrated over the lane of nucleic acids	This may be due to the use of poly(T) tracts in the northern hybridization. To counteract this, include poly(A) at conc. of 1 μg/mL in the hybridization solution
Black spot all over the membrane	This occurs due to the sticking of ^{32}P present as inorganic phosphate or pyrophosphate. Do not use old radiolabeled probes, as hydrolysis can occur in it. In prehybridization/hybridization membrane include 0.5% sodium pyrophosphate

Summary

1. It is a method to identify a target RNA with the help of a probe.
2. In this method, RNA is run in agarose gel to get it separated and then transferred to a suitable membrane with the help of capillary action.
3. The membrane is then fixed and subsequently hybridized with a probe in a hybridization chamber.
4. The signal, due to the binding of the probe to the target RNA fixed to the membrane, is detected by autoradiography.
5. Approximately 20 μg of RNA can be detected in each lane of the gel. Abundant mRNA can be detected by analysis of 10 μg of total cellular RNA, while, in the case of rare RNAs, at least 1.0 μg of poly(A)+ RNA should be loaded in each lane.

CHAPTER 9

Clinical Biochemistry

A. Kidney function test
  1. Estimation of blood urea.
  2. Estimation of blood uric acid.
  3. Estimation of creatinine in serum.
B. Liver function test
  4. Determination of serum glutamate oxaloacetate transaminase (SGOT) activity.
  5. Determination of serum glutamate pyruvate transaminase (SGPT) activity.
  6. Estimation of alkaline phosphatase (ALP) from serum.
  7. Estimation albumin from serum.
C. Diabetes test
  8. Estimation of blood glucose by the glucose oxidase method.
  9. Estimation of urine glucose by Benedict test.
  10. Estimation of urine glucose by Benedict titration method.
  11. Oral glucose tolerance test.
D. Lipid profile test
  12. Estimation of total cholesterol in serum (Watson method).
  13. Estimation of total cholesterol in serum (enzymatic method).
  14. Estimation of triglyceride (TG) in serum.
E. Other clinical test
  15. Hemoglobin estimation in blood.

DETERMINATION OF SERUM UREA [DIACETYL MONOXIME (DAM) METHOD]

Definition

Urea is a nitrogenous waste and formed as a by-product during protein metabolism. The amino group ($-NH_2$) from amino acids converted to ammonia which further enter the ornithine cycle in the liver to be converted into urea. The urea was then transported to the kidney and excreted out along with urine. The normal level of urea in the serum is 10–40 mg/dL. Excretion of urea in urine is 12–24 g/24 h. High serum urea indicates renal dysfunction, urinary tract obstruction, etc.

Rationale

In the strong acidic condition and the presence of ferric ions, urea condenses with diacetyl monoxime and forms a pink-colored complex (a diamine derivative). This pink-colored complex absorbs maximum at 520 nm filter. The absorbance or intensity of the complex is directly proportional to the concentration of urea in the serum. Ferric ion and thiosemicarbazide acts as a catalyst. Thiosemicarbazide intensifies the color, stabilizes it, and avoids deproteinization of serum.

$$\text{Urea} + \text{Diacetyl monoxime} \xrightarrow{\text{Heat, acid, Fe}^{3+}} \underset{\text{(Diamine derivative)}}{\text{Pink color complex}}$$

Thus, the colorimetric estimation of this pink-colored complex at 520 nm is a proportional measure of the concentration of the serum urea.

Materials, equipment, and reagents

A. **Reagents**: Thiosemicarbazide (TSC), ferric chloride, sulfuric acid, phosphoric acid, diacetylmonoxime (DAM), urea standard, sample (serum/urine).
B. **Glassware**: Test tube, test tube holder, dropper, beaker, cuvette, pipette.
C. **Instrument**: Colorimeter/spectrophotometer.

Protocols

1. Take three test tubes and label them as standard (S), blank (B), and test (T).
2. Add the reagent as shown in the table:

Contents in the Test Tube	Blank (B)	Standard (S)	Test (T)
Urea standard (40 mg/dL)	–	0.01	–
Sample	–	–	0.01 mL
Distilled water	0.01 mL	–	–
Acid reagent	1 mL	1 mL	1 mL
DAM/TSC reagent	0.5 mL	0.5 mL	0.5 mL
Mix carefully, boil at 100°C for 10 min			
Cool under a running tap for 5 min			
OD at 520 nm	OD_B	OD_S	OD_T

3. Calculate the amount of urea in the sample.

Calculation

$$\text{Urea (mg/dL)} = \frac{OD_T - OD_B}{OD_S - OD_B} \times \frac{\text{Concentratin of}}{\text{standard (40 mg/dL)}}$$

Protocols in Biochemistry and Clinical Biochemistry. https://doi.org/10.1016/B978-0-12-822007-8.00001-5

Precursor techniques

1. Acid reagent: Take 100 mL of 85% phosphoric acid and 300 mL of conc. sulfuric acid and make final volume 1000 mL by distilled water. Add 100 mg of ferric chloride into it.
2. DAM reagent/TSC reagent: Dissolve 2 g of DAM reagent and 0.4 g of TSC reagent in 100 mL of distilled water. Store at room temperature for 6 months.
3. Urea standard (40 mg/dL): Dissolve 40 mg of urea in 100 mL of distilled water.

Safety considerations and standards

1. Blood hemolysis should be avoided while collecting blood for the analysis.
2. Reading should be taken within 30 min as color is not stable after that.

Analysis and statistics

The amount of urea in the given sample is _____ mg/dL.

Urea nitrogen in mg/dL = Urea in mg/dL × 0.467.

Pros and cons

Pros	Cons
Easy and cheap method	Color is stable for 30 min only
No interference from the ammonia	Time taking process
More accurate and sensitive than the Berthelot method	

Alternative methods/procedures

Berthelot method, Nessler's method.

Summary

1. These are the best and easy methods of urea estimation in the laboratory.
2. This method is based on the principle that urea reacts with diacetyl monoxime at high temperature and in an acidic medium to form a pink-colored diamine derivative. This complex shows absorbance at 520 nm. The absorbance of the solution is a direct measurement of the urea concentration.
3. Ferric ion and thiosemicarbazide acts as a catalyst and stabilizes the color.
4. High serum urea is associated with defective renal functions.

ESTIMATION OF URIC ACID IN SERUM SAMPLE (URICASE METHOD)

Definition

Uric acid is the metabolite formed during the breakdown of purine nucleosides. It is the metabolic waste and excreted out from the kidney as sodium urate. The normal blood uric acid level is 2.5–6 mg/dL in females and 3.4–7 mg/100 mL of blood in males. In urine, its level should be 250–750 mg/dL. The high level of uric acid in serum is the condition of hyperuricemia. Hyperuricaemia may be caused by defective kidney function, gout, leukemia, diabetes, hypothyroidism, etc.

Rationale

Serum uric acid estimation is another parameter that is used along with urea and creatinine for assessing kidney function. In the enzymatic method of uric acid estimation, the uricase enzyme acts upon uric acid and forms allantoin and hydrogen peroxide. The by-product of H_2O_2 in the presence of catalytic enzyme peroxidase reacts with 4-amino antipyrine and dihydroxybenzene sulfonic acid (DHBS) to form a red-colored quinone-imine dye complex. This dye absorbs at 505 nm. The amount of uric acid in serum is directly proportional to the absorbance of the final product.

$$\text{Uric acid} + H_2O_2 + O_2 \xrightarrow{\text{Uricase}} \text{Allantoin} + H_2O_2 + CO_2$$

$$H_2O_2 + \text{4-Aminoantipyrine} + \text{DHBS} \xrightarrow{\text{Peroxidase}} \begin{array}{l}\text{Quinone imine}\\ \text{dye (Red)}\end{array} + H_2O$$

Materials, equipment, and reagents

A. **Chemicals**: Enzymes (uricase, peroxidase), uric acid standard, sample.
B. **Glassware**: Test tube, dropper, beaker, measuring cylinder, cuvette.
C. **Instrument**: Spectrophotometer/colorimeter.

Protocol

1. First, collect the blood and separate the serum.
2. Now use this protein-free filtrate for uric acid estimation. Take three clean test tubes and label them as T (Test), S (standard), and B (Blank).

Reagent	Blank	Standard	Test
Protein free sample	–	–	0.02 mL
Distilled water	0.02 mL	–	–
Uric acid standard	–	0.02 mL	–
Enzyme and buffer reagent	1 mL	1 mL	1 mL
	Mix well and keep it 15 min at room temperature		
Absorbance at 505 nm	OD_B	OD_S	OD_T

Calculations

The concentration of uric acid (mg/100 mL of blood):

$$\text{Uric acid (mg/dL)} = \frac{OD_T - OD_B}{OD_S - OD_B} \times \begin{array}{l}\text{Concentration of}\\ \text{standard (6 mg/dL)}\end{array}$$

Precursor techniques
1. Enzymes reagent and enzyme buffer.
2. Uric acid standard (6 mg/dL): Dissolve 6 mg of uric acid in 100 mL of water.

Safety considerations and standards
1. Handle the acids carefully.
2. Avoid hemolysis of blood while sample preparation.

Pros and cons

Pros	Cons
Easy and reproducible results. Higher precision	Expensive chemicals/enzymes
It can be used for routine use	Time taking process
The enzymatic method is more specific than chemical methods	Interference by some endogenous substance like ascorbic acid

Alternative methods and protocols
The phosphotungstic acid method, oscillating reaction method, HPLC method, titration method.

Summary
1. It is a very easy method of uric acid estimation.
2. It is an enzymatic method. In this method, the uricase enzyme produces hydrogen peroxide as a by-product by oxidizing uric acid. This hydrogen peroxide forms a red dye by reacting with chromogen sulfonic acid and amino antipyrine in the presence of peroxidase. The absorbance at 505 nm is a direct measurement of uric acid in the sample.

ESTIMATION OF CREATININE IN SERUM SAMPLES (JAFFE'S METHOD)
Definition
Creatine is synthesized in the liver and kidney and stored in skeletal muscle as creatine phosphate as storage energy. The creatine is converted to creatinine daily which is excreted out through urine. Increased level of serum or urine creatinine level is an indicator of defective renal function. The normal blood creatinine level is 0.6–1.2 mg/100 mL of blood.

Rationale
In the alkaline conditions, creatinine forms a red-orange complex by reacting with picric acid. The intensity of this colored complex (creatinine-picrate) is directly proportional to the amount of creatinine present in the sample. The complex shows absorption maxima at 520 nm.

$$Creatinine + Picric\ acid \xrightarrow{NaOH} Creatinine\text{-}picrate\ complex$$
$$\text{(Reddish–oragnge)}$$

Materials, equipment, and reagents
A. **Chemicals**: Sodium hydroxide, picric acid, creatinine powder, HCl, H_2SO_4, sodium tungstate, sample.
B. **Glassware**: Test tube, test tube holder, dropper, beaker, measuring cylinder.
C. **Instrument**: Spectrophotometer.

Protocol
1. **Precipitation of protein**: Take 1 mL of serum in a clean test tube, add 3 mL of distilled water, 1 mL of 10% sodium tungstate, 1 mL of 2/3 N H_2SO_4, drop by drop with constant shaking, and mix well. Keep it for 10 min, centrifuge it for 5 min, and take the supernatant in a separate clean test tube.
2. Now use this protein-free filtrate for creatinine estimation. Take three clean test tubes and label them as T (test), S (standard), and B (blank).

Reagent	Blank	Standard	Test
Protein free filtrate	–	–	3 mL
Distilled water	3 mL	–	–
Creatinine powder	–	3 mL	–
Picric acid	1 mL	1 mL	1 mL
NaOH	1 mL	1 mL	1 mL
	Mix well and keep it 15 min at room temperature		
Absorbance at 520 nm	OD_B	OD_S	OD_T

Calculations
The concentration of creatinine (mg/100 mL of blood)

$$\frac{OD_T - OD_B}{OD_S - OD_B} \times \frac{\text{Concentration of the standard}\,(0.01)}{\text{Volume of sample}\,(0.5)} \times 100$$

- The volume of sample: Since we took 1 mL of a serum sample for the first step to remove proteins. The final volume was 6 mL. Out of 6 mL, we used 3 mL for creatinine estimation, thus the used sample volume would be 0.5 mL.

Precursor techniques
1. Picric acid (0.04 M): Dissolve 9.16 g of picric acid in 1 L of distilled water with mild heating.
2. Sodium hydroxide (0.75 N): Dissolve 3 g of NaOH in 100 mL of distilled water.
3. Sodium tungstate (10%): Dissolve 10 g of sodium tungstate in 100 mL of distilled water.

4. Standard creatinine solution (0.01 mg/mL): First dissolve 100 mg of creatinine powder in 100 mL of 0.1 N HCl (1 mg/mL). Dilute it by 100 times which means take 1 mL of it to make the volume 100 mL by distilled water.
5. 2/3 N H_2SO_4: Dissolve 8.7 mL of 97% H_2SO_4 (specific gravity 1.97) in 491.3 mL of distilled water.

Safety considerations and standards
1. Handle the acids carefully.
2. Do pipetting accurately.

Pros and cons

Pros	Cons
Easy and cheap method	Sensitivity is less
It can be used for routine use	Affected by noncreatinine chromogens like bilirubin
	It is less specific. Other components like glucose, urea, uric acid, etc. also contributes to the color development

Alternative methods and protocols
Enzymatic method.

Troubleshooting

Problem	Solution
Color development due to nonspecific chromogens	Add Lloyd's reagent (hydrated aluminum silicate) in acidic condition, centrifuge and resuspend the pellet in alkaline medium and estimate creatinine

Summary
1. It is the most frequently used method for creatinine estimations in the laboratory.
2. Creatinine reacts with alkaline picrate and forms a reddish-orange complex whose absorbance is directly proportional to the amount of creatinine.
3. In the case of creatinine estimation in the urinary sample, the removal of proteins from the sample is not required.

DETERMINATION OF SERUM GLUTAMATE PYRUVATE TRANSAMINASE (SGPT) ACTIVITY (REITMAN AND FRANKEL METHOD, 1957)
Definition
Serum glutamate pyruvate transaminase (SGPT) and serum glutamate oxaloacetate transaminase (SGOT) activity is measured to assay the liver functions. ALT (alanine aminotransferase; formerly GPT—glutamate pyruvate transaminase) and AST (aspartate-aminotransferase; formerly GOT—glutamate-oxaloacetate-transaminase) enzymes are mainly present in liver and kidney cells. ALT is primarily found in the cytoplasm whereas AST is found in both cytoplasm and mitochondria of liver cells. Very less level of the enzymes is detected in the serum but in case of liver damage or injury, their level increases in the blood. These two enzyme levels in the serum are determined to check the liver function.

Rationale
Transaminases are the enzymes which catalyze the transfer of amino group from α-amino acid to α-keto acid. ALT enzyme catalyzes the reaction between L-alanine and α-ketoglutarate to form pyruvate and L-glutamate. The pyruvate reacts with 2,4-dinitrophenylhydrazine in the alkaline condition to form a red-brown complex. The intensity of the color of the complex is directly proportional to the amount of ALT enzyme in the serum.

$$\alpha\text{-Ketoglutarate} + \text{L-alanine} \xrightarrow{\text{ALT}} \text{L-glutamate} + \text{Pyruvate}$$

$$\text{Pyruvate} + 2,4\text{-Dinitrophenylhydrazine} \longrightarrow \begin{array}{c}\text{Brown complex}\\ \text{(Phenylhydrazone)}\end{array}$$

Thus, the colorimetric estimation of this colored complex is a proportional measure of the concentration of the serum alanine transaminase enzymes.

The normal value of SGPT is 6–21 IU (International unit). 1 IU is μmoles of pyruvate formed per minute per L of serum at 37°C.

Materials, equipment, and reagents
A. **Reagents**: Alanine, a-ketoglutarate, NaOH, sample, sodium pyruvate, disodium hydrogen phosphate, potassium dihydrogen phosphate, 2,4-dinitrophenylhydrazine (DNPH).
B. **Glassware**: Test tube, test tube holder, dropper, beaker.
C. **Instrument**: Colorimeter.

Protocols
A. **Preparation of a standard calibration curve**:
   1. Take 1 mL of ALT substrate in six separate test tubes.
   2. Boil it for 5 min at 37°C.
   3. Add 0–0.25 mL of the standard solution in different test tubes.
   4. Make the volume 1.5 mL by adding PBS.
   5. Incubate the mixture at 37°C for 30 min.
   6. Add 0.5 mL of 1 mM DNPH solution in each of the test tubes.

7. Mix well and incubate at room temperature for 15–20 min.
8. Add 5 mL of 0.4 N NaOH.
9. Take the absorbance at 505 nm.
10. Subtract the absorbance value of blank from that of each of the concentrations.
11. Plot the standard calibration curve between final absorbance value (optical density; OD of pyruvate) and volume of the standard solution.

	0	1	2	3	4	5
ALT substrate (mL)	1	1	1	1	1	1
Boiling water bath at 37°C for 5 min						
Standard solution (2-mM sodium pyruvate) (mL)	0	0.05	0.10	0.15	0.20	0.25
Concentration (μmol)	0	0.1	0.2	0.3	0.4	0.5
Phosphate buffer (mL)	0.50	0.45	0.40	0.35	0.30	0.25
Incubate the mixture at 37°C for 30 min						
1 mM DNPH (mL)	1	1	1	1	1	1
Mix and incubate at room temperature for 15–20 min						
0.4 N NaOH	5	5	5	5	5	5
After 10 min absorbance at 505 nm						
Absorbance (505 nm)	A_0	A_1	A_2	A_3	A_4	A_5
Final value	A_0	A_1-A_0	A_2-A_0	A_3-A_0	A_4-A_0	A_5-A_0

B. SGPT assay in the sample:
1. Take 1 mL of ALT substrate in two separate test tubes and label them as B (blank) and T (test). Boil it for 5 min at 37°C.
2. Add 0.2 mL of serum sample in the T-test tube.
3. Incubate the mixture at 37°C for 30 min.
4. Add 1 mL of 2,4-dinitrophenylhydrazine solution, mix, and incubate at room temperature for 15–20 min.
5. Add 0.2 mL of serum in the B test tube.
6. Mix and keep it for 10 min.
7. Add 0.4 mL of NaOH and after 10 min read the absorbance at 505 nm.

Contents in the Test Tube	Blank (B)	Test (T)
ALT substrate	1 mL	1 mL
Boiling water bath at 37°C for 5 min		
Serum	–	0.2 mL
37°C for 30 min		
DNPH	1 mL	1 mL
Room temperature for 15–20 min		
Serum	0.2 mL	–
Mix and keep for 10 min		
0.4 N NaOH	5 mL	5 mL
Mix well and keep for 10 min		
OD at 505 nm	B	T
Difference in OD		T–B

Calculation
1. The value of the OD (difference) corresponds to the concentration of the product (in μmol), i.e., standard sodium pyruvate (from calibration curve).
2. SGPT activity (IU/L)

$$\frac{\text{μmole concentration of the product}}{\text{The volume of the sample}} \times \frac{1000}{\text{Incubation time (30 min)}}$$

Precursor techniques
1. Phosphate buffer (0.1 M) pH 7.4: Mix 420 mL of 0.1 M disodium hydrogen phosphate and 80 mL of 0.1 M potassium dihydrogen phosphate. Dissolve 13.97 g of Na_2HPO_4 (0.1 M) and 2.69 g of KH_2PHO_4 (0.1 M) in 1 L of distilled water and adjust the pH to 7.4.
2. SGPT substrate: 200 mM alanine (1.78 g) and 2 mM α-ketoglutarate (29.2 mg) in 0.1 M phosphate buffer and adjust pH 7.4 using 10% NaOH .
3. 0.4 N NaOH: Dissolve 1.5 g of NaOH in 10 mL of distilled water.
4. 2 mM Pyruvate (standard solution): Dissolve 22.0 mg of sodium pyruvate in 100 mL of phosphate buffer.
5. 1 mM DNPH: Dissolve 19.8 mg of 2,4-dinitrophenylhydrazine in 100 mL of 1 N hydrochloric acid.

Safety considerations and standards
1. Blood hemolysis should be avoided while collecting blood for the analysis.
2. Incubation time should be exact 30 min.

Analysis and statistics
The SGPT activity in the blood is calculated in the international unit in a liter (IU/L). The standard value of the serum ALT is between 6 and 22 IU/L. An elevated level is the indication of liver damage or liver disease.

Pros and cons

Pros	Cons
An easy and efficient method	Interference by easily oxidized species like ascorbic acid, uric acid, etc.
There is no need for protein precipitation, extraction in this method	DNPH has a high background value

Alternative methods/procedures
The spectrophotometric method by Karman et al., fluorimetric assay by Rietz and Guilbault, and radiochemical analysis by Parvin et al.

Summary

1. This is a widely used method for the estimation of alanine transaminase (ALT) activity in the serum/blood.
2. The L-alanine and α-ketoglutarate in the presence of ALT enzyme form pyruvate and L-glutamate. Pyruvate reacts with DNPH to form a brown-colored complex. The absorbance of the colored complex measured at 505 nm using a colorimeter.
3. The first standard calibration curve is plotted using a 2 mM pyruvate solution. Then, the absorbance of the test sample is measured at 505 nm. The concentration of the liberated product, i.e., pyruvate in the sample is calculated. From the concentration of the product, the activity of the SGPT is calculated.
4. Increase the level of serum ALT activity (SGPT) is an indicator of liver dysfunction. SGPT assay is a more specific marker for liver disease than SGOT.

DETERMINATION OF SERUM GLUTAMATE OXALOACETATE TRANSAMINASE (SGOT) ACTIVITY

Rationale

AST enzyme catalyzes the reaction between L-aspartic acid and α-ketoglutarate to form glutamate and oxaloacetate. The oxaloacetate reacts with 2,4-dinitrophenylhydrazine in the alkaline condition to form a red-brown complex. The intensity of the color of the complex is directly proportional to the amount of AST enzyme in the serum.

$$\alpha-\text{Ketoglutarate} + L-\text{Aspartic acid} \xrightarrow{\text{AST}} L-\text{glutamate} + \text{Oxaloacetate}$$

$$\text{Oxaloacetate} + 2,4 - \text{Dinitrophenylhydrazine} \longrightarrow \begin{array}{c}\text{Brown complex}\\ \text{(Phenylhydrazone)}\end{array}$$

Thus, the colorimetric estimation of this colored complex is a proportional measure of the concentration of the serum alanine transaminase enzymes.

The normal value of SGOT is 5–40 IU (International unit). 1 IU is μmoles of oxaloacetate formed per minute per liter of serum at 37°C.

Materials, equipment, and reagents

A. **Reagents**: Aspartic acid, a-ketoglutarate, NaOH, sample, oxaloacetic acid, disodium hydrogen phosphate, potassium dihydrogen phosphate, 2,4-dinitrophenylhydrazine (DNPH).

B. **Glassware**: Test tube, test tube holder, dropper, beaker.

C. **Instrument**: Colorimeter.

Protocols

1. **Preparation of a standard calibration curve**:
   1. Take 1 mL of AST substrate in six separate test tubes.
   2. Boil it for 5 min at 37°C.
   3. Add 0–0.25 mL of the standard solution in different test tubes.
   4. Make the volume 1.5 mL by adding PBS.
   5. Incubate the mixture at 37°C for 30 min.
   6. Add 0.5 mL of 1 mM DNPH solution in each of the test tubes.
   7. Mix well and incubate at room temperature for 15–20 min.
   8. Add 5 mL of 0.4 N NaOH.
   9. Take the absorbance at 505 nm.
   10. Subtract the absorbance value of blank from that of each of the concentrations.
   11. Plot the standard calibration curve between final absorbance value (optical density; OD of pyruvate) and volume of the standard solution.

	0	1	2	3	4	5
AST substrate (mL)	1	1	1	1	1	1
Boiling water bath at 37°C for 5 min						
Standard solution (2 mM oxaloacetate/ sodium pyruvate) (mL)	0	0.05	0.10	0.15	0.20	0.25
Concentration (μmol)	0	0.1	0.2	0.3	0.4	0.5
Phosphate buffer (mL)	0.50	0.45	0.40	0.35	0.30	0.25
Incubate the mixture at 37°C for 60 min						
1 mM DNPH (mL)	1	1	1	1	1	1
Mix and incubate at room temperature for 15–20 min						
0.4 N NaOH	5	5	5	5	5	5
After 10 min absorbance at 505 nm						
Absorbance (505 nm)	A_0	A_1	A_2	A_3	A_4	A_5
Final value	A_0	A_1-A_0	A_2-A_0	A_3-A_0	A_4-A_0	A_5-A_0

2. **SGOT assay in the sample**:

1. Take 1 mL of AST substrate in two separate test tubes and label them as B (blank) and T (test). Boil it for 5 min at 37°C.
2. Add 0.2 mL of serum sample in the T-test tube.
3. Incubate the mixture at 37°C for 60 min.
4. Add 1 mL of 2,4-dinitrophenylhydrazine solution, mix, and incubate at room temperature for 15–20 min.
5. Add 0.2 mL of serum in the B test tube.
6. Mix and keep it for 10 min.
7. Add 0.4 mL of NaOH and after 10 min read the absorbance at 505 nm.

Contents in the Test Tube	Blank (B)	Test (T)
AST substrate	1 mL	1 mL
Boiling water bath at 37°C for 5 min		
Serum	–	0.2 mL
37°C for 60 min		
DNPH	1 mL	1 mL
Room temperature for 15–20 min		
Serum	0.2 mL	–
Mix and keep for 10 min		
0.4 N NaOH	5 mL	5 mL
Mix well and keep for 10 min		
OD at 505 nm	B	T
Difference in OD	T – B	

Calculation

1. The value of the OD (difference) corresponds to the concentration of the product (in μmole), i.e., standard oxaloacetate/sodium pyruvate (from calibration curve).
2. SGOT activity (IU/L)

$$\frac{\mu mole\ concentration\ of\ the\ product}{The\ volume\ of\ the\ sample} \times \frac{1000}{Incubation\ time\ (60\ min)}$$

Precursor techniques

1. Phosphate buffer (0.1 M) pH 7.4: Mix 420 mL of disodium hydrogen phosphate (0.1 M) and 80 mL of potassium dihydrogen phosphate (0.1 M). Dissolve 13.97 g Na_2HPO_4 (0.1 M) and 2.69 g KH_2PHO_4 in 1 L of distilled water and adjust pH to 7.4.
2. SGOT substrate: 200 mM aspartate (2.66 g) and 2 mM a-ketoglutarate (29.2 mg) in 0.1 M phosphate buffer, and adjust pH 7.4 using 10% NaOH.
3. 0.4 N NaOH: 1.5 g of NaOH in 10 mL of distilled water.
4. 2 mM oxaloacetate/pyruvate (standard solution): Dissolve 26.4 mg of oxaloacetate (22 mg sodium pyruvate) in 100 mL of phosphate buffer.

5. 1 mM DNPH: Dissolve 19.8 mg of 2,4-dinitrophenylhydrazine in 100 mL of 1 N hydrochloric acid.

Safety considerations and standards

1. Hemolysis of blood should be avoided.
2. Incubation time should be exact 30 min.
3. High level of aldehydes and ketones results in alleviation of results, and thus blank samples should be run using serum after incubation.

Analysis and statistics

The SGOT activity in the blood is calculated in the international unit in a liter (IU/L). The standard value of the serum ALT is between 5 and 40 IU/L. SGOT level enhances during myocardial infarction, liver disease. SGPT level is higher than SGOT during hepatitis, liver disease, hepatocellular destruction, or damage.

Pros and cons

Pros	Cons
An easy and efficient method	Interference by easily oxidized species like ascorbic acid, uric acid, etc. A high level of serum pyruvate interferes with the results
There is no need for protein precipitation, extraction in this method	DNPH has a high background value

Alternative methods/procedures

The spectrophotometric method by Karman et al., fluorimetric assay by Rietz and Guilbault, and radiochemical analysis by Parvin et al.

Summary

1. This is a widely used method for the estimation of aspartate transaminase (AST) activity in the serum/blood.
2. The aspartic acid and α-ketoglutarate in the presence of AST enzyme form oxaloacetate and L-glutamate. Oxaloacetate reacts with DNPH to form a brown-colored complex. The absorbance of the colored complex measured at 505 nm using a colorimeter.
3. The first standard calibration curve is plotted using a 2 mM pyruvate or oxaloacetate solution. Then, the absorbance of the test sample is measured at 505 nm. The concentration of the liberated product, i.e., pyruvate in the sample is calculated. From the concentration of the product, the activity of the SGPT is calculated.

4. The oxaloacetate formed is converted to pyruvate, and thus sodium pyruvate solution can be taken as standard.
5. SGPT assay is a more specific marker for liver disease than SGOT.

COLORIMETRIC ESTIMATION OF ALKALINE PHOSPHATASE (ALP) FROM SERUM (KING AND ARMSTRONG)

Definition
This estimation is made to measure the amount of alkaline phosphatase (ALP) in the serum. The ALP is mainly present in the liver, bone, and kidney. Any damage or injury in the liver leads to the outflow of ALP from the liver to blood. Thus, the high level of ALP in blood indicates liver disease.

Rationale
Phosphatases are the enzymes which break phosphate monoester bond to liberate free phosphate. When disodium phenyl phosphate is added to the serum sample, alkaline phosphatase in the serum breaks it to form phenol and phosphate radical at alkaline pH (pH 10). The liberated phenol then reacts with 4-amino antipyrine in the presence of potassium ferricyanide (an oxidizing agent) and forms a red-orange to brown-colored complex whose absorbance can be measured by colorimeter at 510 nm.

$$\text{Disodium phenyl phosphate} \xrightarrow[\text{Potassium ferricyanide}]{\text{ALP}} \text{Phenol} + \text{Phosphate}$$

$$\text{Phenol} + 4 - \text{Aminoantipyrine} \rightarrow \text{Red} - \text{orange/Brown complex}$$

The intensity of the color is proportional to the phenol formed which in turn related to the activity of the ALP enzyme.

The normal value of ALP is 3–13 KAU (King and Armstrong unit). 1 KAU is mg of phenol liberated per minute per 100 mL of serum.

1 IU (International unit) of ALP = 7.13 KAU

Materials, equipment, and reagents
A. **Reagents**: Disodium phenyl phosphate, potassium ferricyanide, 4-amino antipyrine, NaOH, Na_2CO_3.
B. **Glassware**: Test tube, test tube holder, dropper, beaker, measuring cylinder.
C. **Instrument**: Colorimeter.

Protocols
1. Take the three test tubes and label them as blank (B), standard (S), and test (T).

2. Add the reagents as shown in the table.

Reagent	Blank	Standard	Test
Buffered substrate	0.5 mL	0.5 mL	0.5 mL
Distilled water	1.55 mL	1.5 mL	1.5 mL
Mix well and incubate at 37°C for 3 min			
Serum sample	–	–	0.05 mL
Standard phenol solution	–	0.05 mL	–
Mix well and incubate at 37°C for 15 min			
NaOH	0.8 mL	0.8 mL	0.8 mL
Sodium bicarbonate (0.5 M)	1.2 mL	1.2 mL	1.2 mL
4-Amino antipyrine	1 mL	1 mL	1 mL
Potassium ferricyanide	1 mL	1 mL	1 mL
Mix and take absorbance at 510 nm	OD_B	OD_S	OD_T

Calculation
The alkaline phosphatase activity is expressed as King and Armstrong unit (KAU). KAU is mg of phenol released or liberated from the 100 mL of serum.

The amount of phenol liberated from 100 mL of serum can be calculated as

$$\frac{OD_T - OD_B}{OD_S - OD_B} \times \frac{\text{Concentration of the standard}\,(10) \times 100}{0.1}$$

Precursor techniques
1. Disodium phenyl phosphate (10 mmol/L): Dissolve 0.218 g of disodium phenyl phosphate in 100 mL of distilled water. Boil to dissolve. Cool and add 0.4 mL of chloroform.
2. Buffer (carbonate-bicarbonate buffer, 100 mM): Dissolve 0.636 g of sodium carbonate and 0.336 g sodium bicarbonate in 100 mL of distilled water.
3. Buffered substrate (pH 10): Mix reagents 1 and 2 in a 1:1 ratio.
4. Standard phenol solution (10 mg/dL): Dissolve 10 mg of crystalline phenol in 100 mM HCl. Store in a brown bottle.
5. Sodium hydroxide (0.5 M): Dissolve 2 g of sodium hydroxide in 100 mL of distilled water.
6. Sodium bicarbonate solution (0.5 M): Dissolve 4.2 g of sodium bicarbonate solution in 100 mL of distilled water.
7. Chromogen 4-amino antipyrine: Dissolve 0.6 g of chromogen 4-amino antipyrine in 100 mL of distilled water.
8. Potassium ferricyanide: Dissolve 2.4 g of potassium ferricyanide in 100 mL of distilled water.

Safety considerations and standards

1. Reagents should be stored in brown bottles.
2. The volume of the reagents should be accurate.

Analysis and statistics

The ALP activity of the given sample is _____ KAU.

Pros and cons

Pros	Cons
Disodium phenyl phosphate hydrolyzed more rapidly than β-glycerophosphate	At high enzyme (ALP) concentrations, its activity toward the substrate is decreased

Alternative methods/procedures

Bodansky method, Huggins, and Talalay.

Summary

1. This is a very good and routinely used method for the determination of the alkaline phosphatase activity.
2. The ALP enzyme will act on the substrate disodium phenyl phosphate and liberate phenol. The liberated phenol is directly proportional to the activity of the ALP enzyme. The liberated phenol level can be detected by treating with chromogen amino antipyrine and potassium ferricyanide. The reaction results in the formation of a red-orange complex whose absorbance can be measured at 510 nm.
3. The amount of phenol liberated by 100 mL of serum is equal to KAU.

ESTIMATION OF SERUM ALBUMIN BY THE DYE-BINDING METHOD (BROMOCRESOL GREEN METHOD)

Definition

Albumin is the most abundant plasma protein (60%) produced in the liver which performs a variety of functions including nutrition, maintenance of osmotic pressure, transport, homeostasis, etc.

Rationale

Albumin in the serum reacts with bromocresol green (BCG), an anionic dye at acidic pH to form a green-blue-colored complex whose absorbance can be measured at 630 nm. The intensity of the color is directly proportional to the albumin level.

$$\text{Albumin} + \text{Bromocresol} \xrightarrow{\text{pH 4.2}} \begin{array}{c}\text{Albumin} - \text{BCG complex} \\ \text{(green} - \text{blue color)}\end{array}$$

Materials, equipment, and reagents

A. **Reagents**: Bromocresol green, albumin.
B. **Glassware**: Test tube, test tube holder, dropper, beaker.
C. **Instrument**: Spectrophotometer.

Protocol

1. Take three clean test tubes and label them as standard (S), test (T), and blank (B).
2. Add 2 mL of water and 1 mL of buffered dye in each of the test tubes.
3. Add 0.01 mL of albumin standard in a standard test tube.
4. Add 0.01 mL of serum sample in the T-labeled test tube.
5. Mix well and take absorbance at 630 nm.

	Standard	Blank	Test
Albumin standard	0.01 mL	–	–
Sample	–	–	0.01 mL
BCG buffered dye	1 mL	1 mL	1 mL
Water	2 mL	2 mL	2 mL
Absorbance (630 nm)	S	B	T
Final absorbance	$\triangle A_{stan} =$ S – B		$\triangle A_{test} =$ T – B

Calculation

Amount of albumin (g/dL):

$$\frac{\triangle A_{test} = T - B}{\triangle A_{stan} = S - B} \times \text{Concentration of standard albumin (5 g/dL)}$$

$\triangle A_{test}$ = Absorbance of test
$\triangle A_{stan}$ = Absorbance of standard
Std. conc. = 5 g/dL

Precursor techniques

1. Albumin standard: Dissolve 5 g of albumin in 10 mL of distilled water. Add 50 mg/dL of sodium azide as a preservative.
2. Bromocresol reagent dye:
   a. Succinate buffer (0.1 N)—Dissolve 11.8 g of succinic acid in 800 mL of water. Adjust the pH to 4.0 with 0.1 N NaOH. Make the volume 1 L.

b. Bromocresol green solution (BCG)—Dissolve 419 mg of BCG in 10 mL of 0.1 N NaOH and make the volume 1 L.

c. Buffered bromocresol green solution—Mix 750 mL of solution A+ 250 mL of solution B. Adjust the pH to 4.2 with 0.1 N NaOH solution. Add 4 mL of Brij-35 solution (30%).

Safety considerations and standards

1. Reagent solutions should be prepared freshly and stored in the refrigerator.
2. Pipetting small volume of solution should be avoided.

Analysis and statistics

The normal value of serum albumin is 3.5–5.5 g/dL. The variation of the albumin level occurs in liver disease, kidney dysfunction, malnutrition, etc. The high level of albumin in serum indicates the condition of dehydration.

Pros and cons

Pros	Cons
The results are 90% accurate	Interference by heparin
An easy and efficient method	Sensitivity is less

Alternative methods/procedures

The immunochemical method, capillary zone electrophoresis, and bromocresol purple method.

Summary

1. Bromocresol green is an anionic dye which is a pH indicator. Its yellow color at pH 3.8 converts to green at pH 3.8–5.4 and blue at above pH 5.4.
2. Albumin in the serum can be estimated using bromocresol green. At acidic pH, albumin forms a color complex with bromocresol green which can be measured at 630 nm. The amount of serum albumin is calculated by standard albumin solutions.
3. Albumins can also be estimated in serum using the biuret reaction after precipitation and separation of serum globulins by sodium sulfate. The supernatant albumin is estimated by biuret reaction. Total protein can be estimated by biuret reaction.

$$Globulin = Total\ protein - Albumin$$

Diabetes test:

Diabetes is a major health concern, especially in older people. Diabetes is a metabolic disease defined as high blood sugar either due to decreased production of insulin hormone or resistance to the action of insulin hormone. Insulin hormone produced by beta cells of the pancreases, which helps in maintaining blood glucose homeostasis. High blood glucose level is associated with other complications such as kidney disease, cardiovascular complications, eye damage, etc.

In the laboratory, various tests are performed to check the level of blood and urine glucose.

ESTIMATION OF BLOOD GLUCOSE BY THE GLUCOSE OXIDASE METHOD

Definition

Glucose is a monosaccharide (aldohexose sugar) which is stored as glycogen (polysaccharide) in the liver and skeletal muscle. The breakdown of polysaccharides in the intestine results in monosaccharides such as glucose, fructose, etc. The absorbed glucose molecules in the blood finally transported to cells. The insulin hormone secreted by the beta cells of pancrease plays an important role in their transport inside the cells. Under decreased production of insulin or insensitivity of the insulin receptors (as in diabetes mellitus), it results in the blood glucose level elevation. Thus, the estimation of blood glucose level is the test for diabetes. The normal glucose level is 80–100 mg/dL under fasting condition and 120–140 mg/dL 2 h after eating (random).

Rationale

The glucose oxidase method is also called Trinder's method. In this method, β-D-glucose in the sample first oxidized to gluconic acid by the enzyme glucose oxidase. Hydrogen peroxide which formed as a by-product coupled with 4-amino antipyrine or 4-aminophenazone and phenol to form a red dye named quinonimine in the presence of peroxidase enzyme. The absorbance of the colored product can be measured at 505 nm and its intensity is directly proportional to the amount of glucose.

$$\beta\text{-D-Glucose} + 2H_2O + O_2 \xrightarrow{\text{Glucose oxidase}} \text{Gluconic acid} + H_2O_2$$

$$H_2O_2 + 4\text{-Aminoantipyrine} + \text{Phenol} \xrightarrow{\text{Peroxidase}} \text{Quinonimine dye} + H_2O$$

Materials, equipment, and reagents

A. **Reagents**: Glucose oxidase, 4-amino antipyrine, phenol, and horseradish peroxidase, glucose standard (100 mg/dL), EDTA, sodium fluoride, sample.

B. **Glassware**: Test tube, pipette, tip box, cuvette.

C. **Instruments/apparatus**: Spectrophotometer, water bath.

Protocols

1. Collect the blood in a vial which contains the anticoagulant such as EDTA (1 mg/mL of blood) and sodium fluoride (10 mg/mL of blood). Separate the serum/plasma from the blood and store it at 2–8°C.
2. Take three cleaned test tubes and label them as S (standard), B (blank), and T (test).
3. Add the reagents as shown in the table:

Reagent	Blank	Standard	Test
Sample	–	–	0.01 mL
Glucose standard	–	0.01 mL	–
Water	0.01 mL	–	–
Enzyme reagent (Glucose reagent)	1 mL	1 mL	1 mL
	Incubate at 37°C for 10 min		
Absorbance at 505 nm (OD$_{505}$)	OD$_B$	OD$_S$	OD$_T$

4. Calculate the amount of glucose.

Calculation

$$\text{Glucose (mg/dL)} = \frac{OD_T - OD_B}{OD_S - OD_B} \times \frac{\text{Concentration of standard}}{(100\,\text{mg/dL})}$$

Safety considerations and standards

1. The blood sample should be stored properly if used later and if fresh samples to be used, and then it is to be taken out with proper handling to avoid any kind of infection.
2. Never pipette blood samples by mouth.
3. The glucose reagents or enzymes should be diluted in the appropriate diluent or phosphate buffer (pH 7).

Analysis and statistics

The amount of glucose in the given sample is _____ mg/dL.

Pros and cons

Pros	Cons
Easy, rapid, and more specific method than chemical methods	Sample collected in sodium fluoride vials cannot be used for the estimation of sodium, uric acid
Very less sample is required	Not suitable for the sample from jaundices neonates
This method shows linearity in absorbance up to 500 mg/dL	

Alternative methods/procedures

Hexokinase method, glucose dehydrogenase method, orthotoluidine method, etc.

Summary

1. The glucose oxidase (GOD-POD) method is an easy and accurate method of glucose estimation.
2. Glucose oxidase enzyme extracted from the fungus *Aspergillus niger*. This enzyme specifically acts on β-D-glucose.
3. Glucose oxidase oxidizes glucose into gluconic acid. Hydrogen peroxide is formed as a by-product which reacts with 4-aminoantipyrene and phenol to form a red-colored compound such as quinonimine. The absorbance of glucose in the sample is measured at 505 nm.

ESTIMATION OF GLUCOSE IN URINE BY BENEDICT METHOD

Definition

For the diabetes test, glucose is also estimated in the urine sample along with a blood sample. While filtration of the blood and formation of the urine in the nephrons, glucose is also filtered along with other waste products and water. The glucose is also absorbed back by the tubules of the nephron. Normally the detectable level of glucose in the urine is below 100 mg/dL (0–0.8 mmol/L). A higher level of glucose in the urine is the indication of higher glucose in the blood (diabetes mellitus).

Rationale

In the laboratory urine, glucose can be estimated by the Benedict test. Benedict test is generally used for the detection of reducing sugar in any sample. This test can also be used for estimating the glucose amount in the urine. Glucose present in the sample reduces the copper sulfate (Cu^{2+}) of the benedict reagent (blue color) in the alkaline medium to the cuprous oxide (Cu^+ red color) and oxidizes itself to gluconic acid. More the amount the glucose in the sample, the more the reduction of copper sulfate, and the disappearance of the blue color. Thus, the glucose amount can be checked by the final color development.

$$\underset{\text{(Blue)}}{\text{Copper sulfate}} + \text{Glucose} + OH^- \rightarrow \text{Gluconic acid} + \underset{\text{(Red ppt)}}{\text{Cuprous oxide}}$$

Materials, equipment, and reagents

A. **Reagents**: Sodium citrate, copper sulfate, sodium carbonate, sample, distilled water.
B. **Glassware**: Test tube, test tube holder, pipette, conical flask, beaker, measuring cylinder.
C. **Instrument/apparatus**: Burner.

Precursor technique

1. Benedict reagent: Dissolve 10 g anhydrous sodium carbonate, 17.3 g sodium citrate, and 1.73 copper sulfate in 100 mL of distilled water.

Protocols

1. Take 5 mL of benedict reagent and add 8–10 drops of the urine sample.
2. Boil the sample and cool down.
3. Observe the color change and check the glucose amount from the table.

Observation

Color of the Mixture	Approximate Amount of Glucose	Conclusion
Blue	No glucose	Glucose absent (−)
Green	0.5% (500 mg/dL)	Trace amount
Greenish brown	1% (1000 mg/dL)	(+)
Yellow	1.5% (1500 mg/d $<$)	(++)
Orange	2% (2000 mg/dl)	(+++)
Brick red	Above 2%	(++++)

Safety considerations and standards

1. Disinfect or sterilize all the apparatus used after the test.

Analysis and statistics

The amount of glucose in the given urine sample is .... mg/dL.

Pros and cons

Pros	Cons
An easy and quick method	Gives approximate quantitation

Alternative methods/procedures

Fehling test, modified Folin Wu method.

Summary

1. This is a quick method used in the laboratory for the estimation of glucose in the urine.
2. This method is based on the principle of the reducing capacity of the glucose. Glucose reduces the copper sulfate to the cuprous oxide in the alkaline condition. Upon reduction, the blue color converts to red color. Thus, based on the amount of glucose color intermediates are formed. From the table, the amount of glucose is estimated.

ESTIMATION OF PERCENTAGE OF GLUCOSE (REDUCING SUGARS) IN URINE (BY BENEDICT TITRATION METHOD)

Definition

The Benedict test only gives the approximate glucose amount in the sample. The further titration-based method gives the exact amount of glucose in the urine sample.

Rationale

In the laboratory urine, glucose can be estimated by the Benedict test. During the qualitative test of reducing sugars by benedict reagent, copper sulfate is reduced to the cuprous oxide (Cu_2O red precipitate) in the alkaline condition. In the quantitative estimation of glucose or reducing the sugar by the Benedict test, the reagent also contains potassium thiocyanate along with copper sulfate. Thus, in the reduction reaction of Cu^{2+} by glucose, a white precipitate of cuprous thiocyanate (CuCNS) is formed instead of a red precipitate of cuprous oxide. The potassium ferricyanide of the benedict reagent keeps the cuprous oxide in the solution. Copper sulfate and sodium carbonate (Na_2CO_3) reacts to form $CuCO_2$ which remains dissolved by the sodium citrate. The formation of white precipitate with complete disappearance of the blue color indicates the complete reduction of copper. Thus, glucose in the sample can be calculated titration.

$$\text{Copper sulfate + Glucose} \atop \text{+Potassium thiocyanate} \rightarrow \text{Gluconic acid + Cuprous thiocyanate} \atop \text{(white ppt)}$$

Materials, equipment, and reagents

A. **Reagents**: Sodium citrate, copper sulfate, sodium carbonate, potassium thiocyanate (KCNS), potassium ferrocyanide, glucose, sample, distilled water.
B. **Glassware**: Pipette, conical flask, beaker, measuring cylinder, burette, burette stand, dropper, volumetric flask, funnel.
C. **Instrument/apparatus**: Burner.

Precursor technique

1. Benedict reagent 1: Dissolve 9 g of copper sulfate in 40 mL of distilled water.
2. Benedict reagent 2: Dissolve 100 g of crystalline sodium carbonate in 300 mL boiling distilled water. Add 100 g of crystalline sodium citrate and 64 g of potassium thiocyanate and boil till a clear solution is obtained. Cool it.

3. Working benedict reagent: Mix reagent 1 and 2 and make the final volume 500 mL by distilled water.
4. Glucose standard: Dissolve 1.25 g glucose powder in 250 mL of distilled water in a volumetric flask.
   Note: 25 mL of benedict solution requires 0.05 g of glucose for complete reduction. Thus, if glucose standard of concentration 1.25 g/250 (0.005 g/mL) is used, then the volume of the standard glucose used would be 10 mL.

Protocols
1. Titration with a standard glucose solution:
   a. Fill the burette with a standard glucose solution.
   b. Take 25 mL of benedict solution in the 250 mL volumetric flask and add 1 g of anhydrous sodium carbonate.
   c. Boil the benedict solution with simultaneously adding glucose solution in it. Note the reading of the burette when all the blue color disappears. This is V1. Repeat the process to get two same readings.
2. Titration with an unknown solution:
   a. Now fill the burette with an unknown sample (urine) and repeat the above procedure.
   b. Repeat for two same readings. This is V2.
3. Calculate the amount of glucose.

Calculation
$$N1V1 = N2V2$$

N1 = Standard glucose concentration (0.005 g/mL).
V1 = Volume of standard glucose used (burette reading).
N2 = Glucose amount in the urine sample.
V2 = Volume of urine sample used (burette reading).

$$N2 = N1V1/V2$$
$$\text{Glucose in 100 mL of urine sample} = \frac{N1V1}{V2} \times 100 \text{ g}$$

Safety considerations and standards
1. Disinfect or sterilize all the apparatus used after the test.

Analysis and statistics
The amount of Glucose in the given sample is _____ mg/dL.

Pros and cons

Pros	Cons
Easy and economical	Time taking

Alternative methods/procedures
Fehling test.

Summary
1. This is a very easy method of urine glucose estimation.
2. In this method, the unknown sample is titrated with a benedict solution. The benedict reagent contains potassium thiocyanate and thus, a white precipitate of cuprous thiocyanate is formed instead of a red precipitate of cuprous oxide. The disappearance of a complete blue color is the endpoint of the titration.
3. Generally, 0.05 g glucose is required for a complete reduction of 25 mL benedict reagent.

ORAL GLUCOSE TOLERANT TEST (OGTT)
Definition
Glucose tolerance tests are performed to check the body's ability to utilize blood glucose. It is a very sensitive method. These tests are to assay the glucose tolerance capacity of the body.
   Glucose tolerance tests can be performed by an oral glucose tolerance test and intravenous glucose tolerance test.

Rationale
A glucose tolerance test is usually performed for diabetes patients to check their capacity to utilize the amount of glucose in the circulation. In these tests' patients with overnight fasting for 12 h, are administered with a specific dose (75 g) of glucose, and the blood glucose level is estimated after 2 h.

Materials, equipment, and reagents
All the reagents, equipment are the same that is required for blood glucose estimation methods like glucose oxidase.

Protocols
1. First, collect the venous blood from the patients (12 h fasting overnight).
2. Urine is also collected for glucose estimation.
3. 75 g of glucose is given orally in 250 mL of water or lemon juice.
4. After 2 h the blood sample is collected again, and glucose level is estimated by any method.
5. Now compare the glucose level with the standard table for result interpretation.

Safety considerations and standards
1. Diabetes patients under insulin treatment should not go through this test.
2. 12 h of fasting is a must.
3. Patients should be aware of the test. The patients should be instructed about dietary requirements before the test. Tobacco and smoking are not allowed to patients.

Analysis and statistics

Time Point	Normal (mg/dL)	Impaired Glucose Tolerance (IGT) (mg/dL)	Diabetes (mg/dL)
Fasting (12h)	<110	110–126	>126
After 2h of glucose administration	<140	140–199	>200

Summary

1. This method is used to screen type 2 diabetes and gestational diabetes.
2. The patient is administered with 75 g of glucose after 12 h of overnight fasting. The blood glucose level is estimated by any method like the glucose oxidase method. Two h after the glucose level is checked again. The blood glucose level is compared with the standard table and accordingly diagnose the disease.
3. In the case of a pregnant lady or gestational diabetes, 100 g of glucose is given, and the glucose level is estimated after 3 h.

Other tests for diabetes:

1. Estimation of glycated hemoglobin (HbA1c).

A small percentage of the hemoglobin in the blood becomes glycosylated by the covalent addition of glucose and its derivatives. Estimation of the glycated hemoglobin (HbA1c) in the blood is used for the diagnosis of diabetic and prediabetic conditions. The glycated Hb estimation provides information about the average level of blood glucose in the last 2–3 months and is usually done in the conjugation of total blood glucose estimation. Normal glycated Hb level in the nondiabetic patient is 4%–5.6%. The level between 5.6% and 6.5% is an indication of a prediabetic zone and above 6.5% is for diabetes.

ESTIMATION OF TOTAL CHOLESTEROL IN SERUM (WATSON METHOD)

Definition

Cholesterol is a steroid, a derived lipid. It is present in the free as well as esterified form. There are several methods of cholesterol estimation based on the color production upon treatment with strong acids or some enzymes.

Cholesterol estimation is performed for checking the lipid profile. Disturbance in the normal lipid profile is an indicator of a diseased state.

Desirable cholesterol level: less than 200 mg/dL.

Moderate risk: 200–240 mg/dL.

High risk or hypercholesterolemia: more than 240 mg/dL.

Rationale

This method of total cholesterol estimation was described by Watson (1960).

Liebermann-Burchard test:

When cholesterol is treated with acetic anhydride, a green-colored complex (a sulfonic acid derivative of cholesterol) is formed. The reaction is carried out in the presence of glacial acetic acid and conc. sulfuric acid. The intensity of the color complex is the direct measurement of the amount of cholesterol present in the sample. This complex shows maximum absorption at 575 nm.

$$\text{Cholesterol} + \text{Acetic anhydride} \xrightarrow[\text{Conc.sulphuric acid}]{\text{Glacial acetic acid}} \begin{array}{l}\text{Green color}\\ \text{complex}\end{array}$$

Materials, equipment, and reagents

A. **Reagents**: 2,5-dimethyl benzene-sulfonic acid, glacial acetic acid, acetic anhydride, conc. sulfuric acid, cholesterol standard, sample.
B. **Glassware**: Test tube, pipette, tip box, cuvette.
C. **Instrument/apparatus**: Spectrophotometer.

Materials, equipment, and reagents

1. 2,5-Dimethyl benzene-sulfonic acid (0.25 M): Dissolve 5.55 g of 2,5-dimethyl benzene-sulfonic acid (DBSA) in 100 mL of glacial acetic acid.
2. Cholesterol reagent 1: Add 300 mL of acetic anhydride, 100 mL glacial acetic acid, and 100 mL of 0.25 M 2,5-dimethyl benzene-sulfonic acid solution. Store it at room temperature.
3. Cholesterol reagent 2: Conc. sulfuric acid.
4. Cholesterol standard (200 mg/dL): Dissolve 200 mg cholesterol in 100 mL of glacial acetic acid.

Protocols

1. First, isolate the blood, and separate the serum. Store the serum at 2–8°C.
2. Take three cleaned test tubes and label them as S (standard), B (blank) and T (test).
3. Add the reagents as shown in the table:

Reagent	Blank	Standard	Test
Sample	–	–	0.2 mL
Cholesterol standard	–	0.2 mL	–
Water	0.2 mL	–	–
Glacial acetic acid	0.2 ml	0.2 mL	0.2 mL
Cholesterol reagent 1	5 mL	5 mL	5 mL
	Cool the mixture		
Cholesterol reagent 2	0.6 mL	0.6 mL	0.6 mL
	Mix it thoroughly till the precipitate dissolve and keep it for 10–20 min at room temperature		

Absorbance at 575 nm (OD$_{575}$)   OD$_B$ ODs   OD$_T$

4. Calculate the amount of cholesterol.

Calculation

$$\text{Cholesterol} \atop \text{(mg/dL)} = \frac{OD_T - OD_B}{OD_T - OD_B} \times \frac{\text{Concentration of standard}}{(200\,\text{mg/dL})}$$

Safety considerations and standards

1. The blood sample should be stored properly if used later and if fresh samples to be used, and then it is to be taken out with proper handling to avoid any kind of infection.
2. Never pipette blood samples by mouth.
3. Glassware should be clean and dry.
4. Acids should be handled carefully. Bottle of acids should be kept tightly stoppered.

Analysis and statistics

The amount of cholesterol in the given sample is _____ mg/dL.

Pros and cons

Pros	Cons
Results are reproducible. Easy method	Use of strong acids
No enzymes are required	The stability of the end colored product varies with acid concentration

Alternative methods/procedures

Zak's method, enzymatic methods, GC-MS method, etc.

Summary

1. This is the chemical method of cholesterol estimation.
2. In this method, cholesterol is treated with acetic anhydride (Liebermann-Burchard test) in the presence of acetic acid and conc. sulfuric acid which results in the formation of the sulfonic acid derivative of cholesterol (green color) whose absorbance is measured at 575 nm.

ESTIMATION OF TOTAL CHOLESTEROL IN SERUM (ENZYMATIC METHOD)

Definition

Cholesterol is a steroid, a derived lipid. It is present in the free as well as esterified form. There are several methods of cholesterol estimation based on the color production upon treatment with strong acids or some enzymes. Use of strong acids and instability of the reagents and products of the chemical methods, the enzymatic method has a certain advantage. This method of total cholesterol estimation was described by Allain et al. (1974).

Rationale

Cholesterol ester is first hydrolyzed into cholesterol and fatty acids by the enzyme cholesterol esterase (CHE). The free cholesterol oxidized to cholest-4-en-3-one and hydrogen peroxide by the enzyme cholesterol oxidase (CHO). The by-product hydrogen peroxide in the presence of peroxidase reacts with phenol and 4-amino antipyrine to produce quinonimine dye which is red color (Trinders reaction). The intensity of the color dye or solution is directly proportional to the amount of cholesterol in the sample. The absorbance can be measured at 505 nm.

$$\text{Cholesterol esters} \xrightarrow{\text{Cholesterol esterase}} \text{Cholesterol + Fatty acids}$$

$$\text{Cholesterol} \xrightarrow{\text{Cholesterol oxidase}} \text{Cholest} - 4 - \text{en} - 3\,\text{one} + H_2O_2$$

$$H_2O_2 + 4 - \text{Aminoantipyrine} \xrightarrow{\text{Peroxidase}} \text{Quinonimine dye}$$
$$+\text{Phenol} \qquad\qquad\qquad\qquad +H_2O$$

Materials, equipment, and reagents

A. **Reagents**: Cholesterol esterase, cholesterol oxidase, 4-amino antipyrine, phenol, and horseradish peroxidase, cholesterol standard (200 mg/dL), sample.
B. **Glassware**: Test tube, pipette, tip box, cuvette.
C. **Instrument/apparatus**: Spectrophotometer.

Protocols

1. Take three cleaned test tubes and label them as S (standard), B (blank), and T (test).

2. Add the reagents as shown in the table:

Reagent	Blank	Standard	Test
Sample	–	–	0.2 mL
Cholesterol standard	–	0.2 mL	–
Water	0.2 mL	–	–
Enzyme reagent	1 mL	1 mL	1 mL
	Incubate at 37°C for 5 min		
Absorbance at 505 nm (OD$_{505}$)	OD$_B$	OD$_S$	OD$_T$

3. Calculate the amount of cholesterol.

Calculation

$$\text{Cholesterol (mg/dL)} = \frac{OD_T - OD_B}{OD_T - OD_B} \times 200$$

Safety considerations and standards

1. The blood sample should be stored properly if used later and if fresh samples to be used, and then it is to be taken out with proper handling to avoid any kind of infection.
2. Never pipette blood samples by mouth.
3. The area should be cleaned after the procedure.

Analysis and statistics

The amount of cholesterol in the given sample is _____ mg/dL.

Pros and cons

Pros	Cons
Easy, rapid, and accurate method	This is not a specific method as some other steroids also oxidized by cholesterol oxidase
Few reagents are used	Sometimes interference by Hb, bilirubin, etc.
It can be used for unesterified cholesterol also	

Alternative methods/procedures

Zak's method, chemical methods, GC-MS method, etc.

Summary

1. This is the enzymatic method of cholesterol estimation.
2. In this enzymatic method, esterified cholesterol is first hydrolyzed, then oxidized to produce hydrogen peroxide which then reacts with phenol and 4-amino antipyrine and form a red-coloredquinonimine dye. The absorbance of the dye is measured at 505 nm to detect the amount of cholesterol in the sample.

ESTIMATION OF TRIGLYCERIDES IN SERUM (ENZYMATIC METHOD)

Definition

Triglycerides are simple lipid which constituted one molecule of glycerol and three molecules of fatty acids. Triglyceride is the storage form of lipid , which is used for energy production. Triglycerides are found circulating in the blood where they are transported by very-low-density lipoprotein (VLDL). Triglycerides level is often estimated as lipid profiling. The elevated level of triglycerides in the blood is termed as hypertriglyceridemia. When both the level of cholesterol and triglycerides in the blood is higher than the desired, this condition is called hyperlipidemia. High triglyceride in the blood is associated with several risks as heart disease.

   Desirable triglyceride level: Less than 150 mg/dL.

Borderline high: 150–199 mg/dL.
High: 200–499 mg/dL.
Very high: more than 499 mg/dL.

Rationale

Triglycerides hydrolyzed into one molecule of glycerol and three molecules of fatty acids by the enzyme lipase. Glycerol is phosphorylated to glycerol-3-phosphate by the enzyme glycerol kinase. The phosphate group is provided by the ATP. The glycerol-3-phosphate is oxidized to dihydroxyacetone phosphate by the enzyme glycerol-3-phosphate oxidase (GPO). Hydrogen peroxide which is formed as a by-product in the above reaction and then condensed with 4-amino antipyrine (4-AAP)/4-aminophenazone (chromogenic substrate) and 4-chlorophenol to form a red-colored quinone-imine dye. The absorbance can be measured at 505 nm and its intensity is directly proportional to the content of triglyceride in the sample.

$$\text{Triglycerides} + H_2O \xrightarrow{\text{Lipoprotein lipase}} \begin{array}{l}\text{Glycerol}\\+3\,\text{molecules of fatty acids}\end{array}$$

$$\text{Glycerol} + \text{ATP} \xrightarrow{\text{Glycerol kinase}} \text{Glycerol} - 3 - \text{phosphate}$$

$$\begin{array}{l}\text{Glycerol} - 3 -\\ \text{phosphate}\\ + H_2O_2\end{array} + O_2 \xrightarrow[\text{oxidase}]{\text{Glycerol-3-phosphatem}} \begin{array}{l}\text{Dihydroxyacetone}\\ \text{phosphate}\end{array}$$

$$\begin{array}{l}H_2O_2 + 4 - \text{Aminoantipyrine}\\ +4 - \text{Chlorophenol}\end{array} \xrightarrow{\text{Peroxidase}} \text{Quinoneimine dye} + H_2O$$

Materials, equipment, and reagents

A. **Reagents**: Working reagent (it contains lipoprotein lipase, ATP, glycerol kinase, glycerol-3-phosphate oxidase, peroxidase, 4-AAP/4-aminophenazone and 4-chlorophenol), triglyceride standard (200 mg/dL), sample.
B. **Glassware**: Test tube, pipettes, tip box, cuvette.
C. **Instruments/apparatus**: Spectrophotometer.

Protocols

1. First, isolate the blood and separate the serum. Store the serum at 2–8°C.
2. Take three cleaned test tubes and label them as S (standard), B (blank), and T (test).
3. Add the reagents as shown in the table:

Reagent	Blank	Standard	Test
Sample	–	–	0.01 mL
Triglyceride standard	–	0.01 mL	–
Water	0.01 mL	–	–

Working reagent	1 mL	1 mL	1 mL
	Mix it keep it for 10 min at room temperature		
Absorbance at 505 nm (OD$_{505}$)	OD$_B$	ODs	OD$_T$

4. Calculate the amount of cholesterol.

Calculation

$$\text{Triglycerides (mg/dL)} = \frac{OD_T - OD_B}{OD_S - OD_B} \times \text{Concentratio of standard (200 mg/dL)}$$

Safety considerations and standards
1. Avoid hemolysis of the blood.
2. Do not use citrate, fluoride, and oxalate while collecting the sample.

Analysis and statistics
The amount of triglycerides in the given sample is _____ mg/dL.

Pros and cons

Pros	Cons
Easy and rapid method	Interference on the action of lipase by the detergents and some drugs
No effect of low bilirubin	Expensive method
Specificity is more than chemical methods	

Alternative methods/procedures
Chemical method. Liquid-phase partition method, GC-MS method, etc.

Summary
1. The serum should be collected after 12 h of fasting.
2. This is the enzymatic method of triglycerides estimation.
3. In this method, the lipase enzyme hydrolyzes the triglyceride into glycerol and fatty acids. Glycerol undergoes phosphorylation then oxidation reaction in which peroxide is generated, which condenses with chromogen 4-amino antipyrine (4-AAP)/4-aminophenazone to form red-colored quinone-imine dye. The absorbance is measured at 505 nm.

ESTIMATION OF HEMOGLOBIN IN THE BLOOD (SAHLI'S METHOD)

Definition
Hemoglobin (Hb) is a complex protein (metalloprotein; a complex of globin protein with heme with iron) present in the RBC. The main function of the Hb is binding with oxygen in the blood and its transport to different parts of the body. The normal range of Hb in the blood is 14–18 g/dL in males and 12–16 g/dL in females. Hemoglobin estimation frequently used lab tests in many diseased conditions as anemia.

Rationale
In this method, an instrument called Sahli hemoglobin-ometer is used, which consists of solid glass with color standard and a calibrated graduated cylinder. In Sahli's method, hemoglobin is first converted to acid hematin by treating with N/10 HCl. Acid hematin is brown. The color intensity of the solution depends upon the concentration of Hb in the blood. The resultant color solution is diluted with water and matched against the color standard on the glass. The reading is in g/dL (g in 100 mL of blood) or g%.

Materials, equipment, and reagents
A. **Reagents**: N/10 HCl, spirit, sample, distilled water.
B. **Glassware**: Test tube, cotton.
C. **Instrument/apparatus**: Sahli's hemoglobinometer.
   Sahli's hemoglobinometer: It consists of
1. **Comparator**: It is a plastic box of a rectangle shape. In the middle, it has a space for calibrated Hb tubes. On either side, it consists of a fixed brown glass tube (standard tube) for matching the color of the Hb tube. These tubes consist of a standard acid hematin solution.
2. **Hb tube**: It is a calibrated tube for Hb (g/100 mL of blood) (2%–24% on one side and 20%–50% Hb on another side).
3. **Hb pipette**: It is a capillary pipette with a marking of 0.02 mL (20 mm).
4. **Glass rod for stirring.**
5. **Dropper.**

Protocols
1. First, clean the apparatus well and dry it before use.
2. Add N/10 HCl in the Hb tube up to the mark 2% or 2 g with the help of a dropper.
3. Clean the finger with spirit and prick it with a needle. Wipe off the starting blood and pipette out 0.02 mL blood in the Hb pipette and immediately mix it into the acid present in the calibrated Hb tube. Mix it well.
4. After expelling all the blood, draw some distilled water in the pipette and mix it in the acid-blood mix of the Hb tube, so that all the blood is transferred to the tube.
5. Now stir the acid hematin solution by glass rode and allow it to stand for 10 min.
6. Dilute the acid hematin solution by adding distilled water to dropwise and stirring by glass rode. Match

the color of the solution in the Hb tube with the color of the standard tube.

7. Repeat the step of dilution and stirring till the color of the Hb tube matches with the standard tube color and note the reading of the Hb tube.

Precursor techniques

1. N/10 (0.1 N HCl): In the laboratory, the HCl is 37% which is around 12 N or 12 M.

   This HCl of 12 N can be diluted to 0.1 N. Add 8.3 mL of 37% HCl in 1000 mL distilled water to make final volume. Acid should be added in water during dilution.

Safety considerations and standards

1. Clean all the parts of the hemoglobinometer by distilled water or some organic solvent before use.
2. Clean the Hb pipette by N/10 HCl.
3. After pricking the finger, clean it with spirit to avoid infection.
4. Don't add the blood that sticks to the outer surface of the pipette.
5. After pipetting add the blood immediately to the acid solution to prevent clotting.
6. While taking reading, the comparator must be at eye level.

Analysis and statistics

The level of Hb in the given sample of blood is _____ g/dL.

Pros and cons

Pros	Cons
Very easy, quick, and inexpensive method	Accuracy is less. More chances of variations
Does not require technical expertise and big lab infrastructure	Time taking as the color of acid hematin develops slowly

Reagents used are nontoxic (as no use of cyanide)

Color is not stable for a longer time hence the reading should be noted exactly when the color development is maximum

Some other forms of hemoglobin such as carboxyhemoglobin and methemoglobin are not converted to acid hematin thus reading is not accurate

Alternative methods/procedures

Cyanmethemoglobin method, Vanzetti's azide methemoglobin, noninvasive method.

Summary

1. This is the most common method for Hb estimation.
2. Hemoglobin comprises heme and protein part globin. Heme is a nonprotein part where iron complex with porphyrin ring. In the heme, the iron present in the ferrous state. Upon treating the Hb with an acid solution, it dissociates into heme and globin. The heme part converts to hematin as ferrous is oxidized to ferric iron. This acid hematin is dark brown.
3. In this method, acid is added to the blood sample and acid hematin is formed which is diluted to match with the standard color of acid hematin.
4. Although this method does not give the actual value but still very common and popular for graduate and postgraduate students to learn and perform.

References

Al-Ani, M., Opara, U., Al-Bahri, D., Al Rahbi, M., 2007. Spectrophotometric quantification of ascorbic acid contents of fruit and vegetables using the 2,4-dinitrophenylhydrazine method. J. Food Agric. Environ. 5, 165–168.

Allain, C.C., Poon, L.S., Chan, C.S., Richmond, W., Fu, P.C., 1974. Enzymatic determination of total serum cholesterol. Clin Chem. 20 (4), 470 475. https://doi.org/10.1093/clinchem/20.4.470.

Ambade, V.N., Sharma, Y., Somani, B., 1998. Methods for estimation of blood glucose: a comparative evaluation. Med. J. Armed Forces India 54, 131–133. https://doi.org/10.1016/S0377-1237(17)30502-6.

Bates, C.J., 1997. Vitamin analysis. Ann. Clin. Biochem. 34, 599–626. https://doi.org/10.1177/000456329703400604.

Benedict, S.R., 1911. The detection and estimation of glucose in urine. J. Am. Med. Assoc. LVII, 1193–1194. https://doi.org/10.1001/jama.1911.04260100019005.

Bligh, E.G., Dyer, W.J., 1959. A rapid method of total lipid extraction and purification. Can. J. Biochem. Physiol. 37, 911–917. https://doi.org/10.1139/o59-099.

Block, W.D., Geib, N.C., 1947. An enzymatic method for the determination of uric acid in whole blood. J. Biol. Chem. 168, 747–756.

Bonsnes, R.W., Taussky, H.A., 1945. On the colorimetric determination of creatinine by the Jaffe reaction. J. Biol. Chem. 158, 581–591.

Bradford, M.M., 1976. A rapid and sensitive method for the quantitation of microgram quantities of protein utilizing the principle of protein-dye binding. Anal. Biochem. 72, 248–254. https://doi.org/10.1006/abio.1976.9999.

Buckley, G.C., Cutler, J.M., Little, J.A., 1966. Serum triglyceride: method of estimation and levels in normal humans. Can. Med. Assoc. J. 94, 886–888.

Budowski, P., Bondi, A., 1957. Determination of vitamin A by conversion to anhydrovitamin A. Analyst 980, 751.

Busher, J.T., 1990. Serum albumin and globulin. In: Walker, H. K., Hall, W.D., Hurst, J.W. (Eds.), Clinical Methods: The History, Physical, and Laboratory Examinations. Butterworths, Boston.

Carr, M.H., Frank, H.A., 1956. Improved method for determination of calcium and magnesium in biologic fluids by EDTA titration. Am. J. Clin. Pathol. 26, 1157–1168. https://doi.org/10.1093/ajcp/26.10.1157.

Carroll, N.V., Longley, R.W., Roe, J.H., 1956. The determination of glycogen in liver and muscle by use of anthrone reagent. J. Biol. Chem. 220, 583–593.

Chaudhary, R., Dubey, A., Sonker, A., 2017. Techniques used for the screening of hemoglobin levels in blood donors: current insights and future directions. J. Blood Med. 8, 75–88. https://doi.org/10.2147/JBM.S103788.

Chaudhuri, R.K., Mukherjee, M., Sengupta, D., Mazumder, S., 2006. Limitation of glucose oxidase method of glucose estimation in jaundiced neonates. Indian J. Exp. Biol. 44, 254–255.

Claude, J.R., 1962. Determination of total serum cholesterol by Watson's method. Modification of the original technic. Pathol. Biol. 10, 1599–1600.

Clegg, K.M., 1956. The application of the anthrone reagent to the estimation of starch in cereals. J. Sci. Food Agric. 7, 40–44.

Cochran, B., Lunday, D., Miskevich, F., 2008. Kinetic analysis of amylase using quantitative benedict's and iodine starch reagents. J. Chem. Educ. 85, 401. https://doi.org/10.1021/ed085p401.

Corso, G., Papagni, F., Gelzo, M., Gallo, M., Barone, R., Graf, M., Scarpato, N., Dello Russo, A., 2015. Development and validation of an enzymatic method for total cholesterol analysis using whole blood spot. J. Clin. Lab. Anal. 30, 517–523. https://doi.org/10.1002/jcla.21890.

de Koning, A.J., 1974. Analysis of egg lipids. A student project. J. Chem. Educ. 51, 48. https://doi.org/10.1021/ed051p48.

Emmerie, A., Engel, C., 1938. Colorimetric determination of α-tocopherol (vitamin E). Recl. Trav. Chim. Pays-Bas 57, 1351–1355. https://doi.org/10.1002/recl.19380571207.

Evans, R.T., 1968. Manual and automated methods for measuring urea based on a modification of its reaction with diacetyl monoxime and thiosemicarbazide. J. Clin. Pathol. 21, 527–529. https://doi.org/10.1136/jcp.21.4.527.

Folch, J., Ascoli, I., Lees, M., Meath, J.A., Lebaron, N., 1951. Preparation of lipide extracts from brain tissue. J. Biol. Chem. 191, 833–841.

Folch, J., Lees, M., Sloane Stanley, G.H., 1957. A simple method for the isolation and purification of total lipides from animal tissues. J. Biol. Chem. 226, 497–509.

Folin, O., Berglund, H., 1922. A colorimetric method for the determination of sugars in normal human urine. J. Biol. Chem. 51, 209–211.

Foulger, J.H., 1931. The user of the molisch (α-naphthol) reactions in the study of sugars in biological fluids. J. Biol. Chem. 92, 345–353.

Fujiwara, M., Matsui, K., 1953. Determination of thiamine by thiochrome reaction. Anal. Chem. 25, 810–812. https://doi.org/10.1021/ac60077a040.

Gallagher, S., Chakavarti, D., 2008. Staining proteins in gels. J. Vis. Exp. https://doi.org/10.3791/760.

Goldring, J.P.D., 2015. Spectrophotometric methods to determine protein concentration. Methods Mol. Biol. 1312, 41–47. https://doi.org/10.1007/978-1-4939-2694-7_7.

Gornall, A.G., Bardawill, C.J., David, M.M., 1949. Determination of serum proteins by means of the biuret reaction. J. Biol. Chem. 177, 751–766.

Goswami, D., Kalita, H., 2013. Rapid determination of iron in water by modified thiocyanate method. Def. Sci. J. 38. https://doi.org/10.14429/dsj.38.4835.

Hammond, J.B., Kruger, N.J., 1988. The Bradford method for protein quantitation. Methods Mol. Biol. 3, 25–32. https://doi.org/10.1385/0-89603-126-8:25.

Hayashi, K., Sasaki, Y., Tagashira, S., Harada, K., Okamura, K., 1978. Spectrophotometric determination of iron (III) with thiocyanate and nonionic surfactant. Bunseki Kagaku 27, 338–343. https://doi.org/10.2116/bunsekikagaku.27.6_338.

Hoch, H., 1943. Micro-method for estimating vitamin A by the Carr-Price reaction. Biochem. J. 37, 425–429. https://doi.org/10.1042/bj0370425.

Huang, X.-J., Choi, Y.-K., Im, H.-S., Yarimaga, O., Yoon, E., Kim, H.-S., 2006. Aspartate aminotransferase (AST/GOT) and alanine aminotransferase (ALT/GPT) detection techniques. Sensors 6, 756–782.

Ichimura, K., Hisamatsu, T., 1999. Effects of continuous treatment with sucrose on the vase life, soluble carbohydrate concentrations, and ethylene production of cut snapdragon flowers. J. Jpn. Soc. Hort. Sci. 68, 61–66.

Jargar, J.G., Hattiwale, S.H., Das, S., Dhundasi, S.A., Das, K.K., 2012. A modified simple method for determination of serum α-tocopherol (vitamin E). J. Basic Clin. Physiol. Pharmacol. 23, 45–48. https://doi.org/10.1515/jbcpp-2011-0033.

Kapur, A., Hasković, A., Klepo, L., Topcagic, A., Tahirovic, I., Sofić, E., 2012. Spectrophotometric analysis of total ascorbic acid content in various fruits and vegetables. Bull. Chem. Technol. Bosnia Herzeg. 38, 39–42.

Keyser, J.W., Spillane, M.T., 1963. Estimation of blood glucose by glucose oxidase methods: the effect of Fluoride. Proc. Assoc. Clin. Biochem. 2, 217–218. https://doi.org/10.1177/036985646300200814.

Keyser, J.W., Stephens, B.T., 1962. Estimation of serum albumin: a comparison of three methods. Clin. Chem. 8, 526–529. https://doi.org/10.1093/clinchem/8.5.526.

King, E.J., 1932. The colorimetric determination of phosphorus. Biochem. J. 26, 292–297.

Kirby, K.S., 1957. A new method for the isolation of deoxyribonucleic acids: evidence on the nature of bonds between deoxyribonucleic acid and protein. Biochem. J. 66, 495–504.

Kitamura, M., Iuchi, I., 1959. An improved diacetylmonoxime method for the determination of urea in blood and urine. Clin. Chim. Acta 4, 701–706. https://doi.org/10.1016/0009-8981(59)90013-0.

Kruger, N.J., 1994. The Bradford method for protein quantitation. Methods Mol. Biol. 32, 9–15. https://doi.org/10.1385/0-89603-268-X:9.

Kumar, D., Banerjee, D., 2017. Methods of albumin estimation in clinical biochemistry: past, present, and future. Clin. Chim. Acta 469, 150–160. https://doi.org/10.1016/j.cca.2017.04.007.

Kumar, V., Pattabiraman, T.N., 1997. Standardization of a colorimetric method for the determination of fructose using o-cresol: sulphuric acid reagent. Indian J. Clin. Biochem. 12, 95–99. https://doi.org/10.1007/BF02867965.

Ledoux, M., Lamy, F., 1986. Determination of proteins and sulfobetaine with the folin-phenol reagent. Anal. Biochem. 157, 28–31. https://doi.org/10.1016/0003-2697(86)90191-0.

Leiboff, S.L., 1928. A colorimetric method for the determination of inorganic phosphate in blood serum. J. Biol. Chem. 79, 611–619. http://www.jbc.org/content/79/2/611.citation.

Li, L.-H., Dutkiewicz, E.P., Huang, Y.-C., Zhou, H.-B., Hsu, C.-C., 2019. Analytical methods for cholesterol quantification. J. Food Drug Anal. 27, 375–386. https://doi.org/10.1016/j.jfda.2018.09.001.

Lowry, O.H., Rosebrough, N.J., Farr, A.L., Randall, R.J., 1951. Protein measurement with the Folin phenol reagent. J. Biol. Chem. 193, 265–275.

Lugosi, R., Thibert, R.J., Holland, W.J., Lam, L.K., 1972. A study of the reaction of urea with diacetyl monoxime and diacetyl. Clin. Biochem. 5, 171–181. https://doi.org/10.1016/s0009-9120(72)80028-6.

Maizel, J.V., 2000. SDS polyacrylamide gel electrophoresis. Trends Biochem. Sci. 25, 590–592. https://doi.org/10.1016/s0968-0004(00)01693-5.

Maness, N., 2010. Extraction and analysis of soluble carbohydrates. Methods Mol. Biol. 639, 341–370. https://doi.org/10.1007/978-1-60761-702-0_22.

Martinek, R.G., 1964. Method for the determination of vitamin E (total tocopherols) in serum. Clin. Chem. 10, 1078–1086.

McMaster, G.K., Carmichael, G.G., 1977. Analysis of single- and double-stranded nucleic acids on polyacrylamide and agarose gels by using glyoxal and acridine orange. Proc. Natl. Acad. Sci. U.S.A 74 (11), 4835–4838. https://doi.org/10.1073/pnas.74.11.4835 PMID: 73185; PMCID: PMC432050.

Momose, T., Ohkura, Y., Tomita, J., 1965. Determination of urea in blood and urine with diacetyl monoxime-glucuronolactone reagent. Clin. Chem. 11, 113–121. https://doi.org/10.1093/clinchem/11.2.113.

Morin, L.G., 1974. Direct colorimetric determination of serum calcium with o-cresolphthalein complexon. Am. J. Clin. Pathol. 61, 114–117. https://doi.org/10.1093/ajcp/61.1.114.

Murmur, J., 1961. A procedure for the isolation of deoxyribonucleic acid from micro-organisms. J. Mol. Biol. 3, 208.

Murray, M.G., Thompson, W.F., 1980. Rapid isolation of high molecular weight plant DNA. Nucleic Acids Res. 8, 4321–4325. https://doi.org/10.1093/nar/8.19.4321.

Nath, M.C., Chakravorty, M.K., Chowdhury, S.R., 1946. Liebermann-Burchard reaction for steroids. Nature 157, 103. https://doi.org/10.1038/157103b0.

Nielsen, S., 2010a. Complexometric determination of calcium. In: Food Analysis Laboratory Manual. ISBN 978-1-4419-1477-4, pp. 61–67.

Nielsen, S., 2010b. Vitamin C determination by indophenol method. In: Food Analysis Laboratory Manual. ISBN 978-1-4419-1477-4, pp. 55–60.

Ntailianas, H.A., Whitney, R.M.L., 1963. Direct complexometric determination of the calcium and magnesium in milk (Eriochromeblack T Method). J. Dairy Sci. 46, 1335–1341. https://doi.org/10.3168/jds.S0022-0302(63)89277-2.

Olson, B.J.S.C., Markwell, J., 2007. Assays for determination of protein concentration. Curr. Protoc. Pharmacol. https://doi.org/10.1002/0471141755.pha03as38. Appendix 3, 3A.

Paraskevopoulou, A., Kiosseoglou, V., 2006. Cholesterol and other lipid extraction from egg yolk using organic solvents: effects on functional properties of yolk. J. Food Sci. 59, 766–768. https://doi.org/10.1111/j.1365-2621.1994.tb08123.x.

Parrish, D., Bauernfeind, J., 2009. Determination of vitamin a in foods—a review. CRC Crit. Rev. Food Sci. Nutr. 9, 375–394 https://doi.org/10.1080/10408397709527240.

Peake, M.J., Pejakovic, M., White, G.H., 1988. Quantitative method for determining serum alkaline phosphatase isoenzyme activity: estimation of intestinal component. J. Clin. Pathol. 41, 202–206.

Rahmatullah, M., Boyde, T.R., 1980. Improvements in the determination of urea using diacetyl monoxime; methods with and without deproteinisation. Clin. Chim. Acta 107, 3–9. https://doi.org/10.1016/0009-8981(80)90407-6.

Rani, J., Raju, D.S.S.K., 2016. Estimation of serum glutamic oxaloacetic transaminase, serum glutamic-pyruvic transaminase, gamma-glutamyl transferase and cholesterol levels in prolonged (30 years) daily consumption coffee in people. Int. J. Res. Med. Sci. 4, 1564–1573.

Reed, K.C., Mann, D.A., 1985. Rapid transfer of DNA from agarose gels to nylon membranes. Nucleic Acids Res. 13, 7207–7221. https://doi.org/10.1093/nar/13.20.7207.

Reitman, S., Frankel, S., 1957. A colorimetric method for the determination of serum glutamic oxalacetic and glutamic pyruvic transaminases. Am. J. Clin. Pathol. 28, 56–63. https://doi.org/10.1093/ajcp/28.1.56.

Rej, J., 1984. Measurement of aminotransferases: part 1. aspartate aminotransferase. Crit. Rev. Clin. Lab. Sci. 21, 99–186. https://doi.org/10.3109/10408368409167137.

Rosenthal, H.L., 1955. Determination of urea in blood and urine with diacetyl monoxime. Anal. Chem. 27, 1980–1982. https://doi.org/10.1021/ac60108a039.

Roy, S., Kumar, V., 2014. A practical approach on SDS PAGE for separation of protein. Int. J. Sci. Res. 3 (8), 955–960.

Ryan, M.A., Ingle, J.D., 1980. Fluorometric reaction rate method for the determination of thiamine. Anal. Chem. 52, 2177–2184. https://doi.org/10.1021/ac50063a042.

Sardesai, V.M., Manning, J.A., 1968. The determination of triglycerides in plasma and tissues. Clin. Chem. 14, 156–161. https://doi.org/10.1093/clinchem/14.2.156.

Sargent, M.G., 1987. Fiftyfold amplification of the Lowry protein assay. Anal. Biochem. 163, 476–481. https://doi.org/10.1016/0003-2697(87)90251-x.

Shah, T., 2016. Utility of osazone test to identify sugars. J. Med. Sci. Clin. Res. 4, 14361–14365.

Shukla, M., Arya, S., 2018. Determination of chloride ion(Cl-) concentration in Ganga river water by Mohr method at Kanpur, India. Green Chem. Technol. Lett. 4, 06–08. https://doi.org/10.18510/gctl.2018.412.

Simpson, R.J., 2010. CTAB-PAGE. Cold Spring Harb Protoc 2010. https://doi.org/10.1101/pdb.prot5412. pdb.prot5412.

Smith, D., Paulsen, G.M., Raguse, C.A., 1964. Extraction of total available carbohydrates from grass and legume tissue. Plant Physiol. 39, 960–962. https://doi.org/10.1104/pp.39.6.960.

Smith, P.K., Krohn, R.I., Hermanson, G.T., Mallia, A.K., Gartner, F.H., Provenzano, M.D., Fujimoto, E.K., Goeke, N.M., Olson, B.J., Klenk, D.C., 1985. Measurement of protein using bicinchoninic acid. Anal. Biochem 150, 76–85. https://doi.org/10.1016/0003-2697(85)90442-7.

Sobel, A.E., Werbin, H., 1945. Spectrophotometric study of a new colorimetric reaction of vitamin A. J. Biol. Chem. 159, 681–691.

Stepka, W., 1957. Identification of amino acids by paper chromatography. Methods Enzymol. 3, 504–528. https://doi.org/10.1016/S0076-6879(57)03420-5.

Stern, J., Lewis, W.H., 1957. The colorimetric estimation of calcium in serum with ocresolphthalein complexone. Clin. Chim. Acta 2, 576–580. https://doi.org/10.1016/0009-8981(57)90063-3.

Sullivan, D.R., Kruijswijk, Z., West, C.E., Kohlmeier, M., Katan, M.B., 1985. Determination of serum triglycerides by an accurate enzymatic method not affected by free glycerol. Clin. Chem. 31, 1227–1228.

Sur, B.K., Shukla, R.K., Agashe, V.S., 1972. The role of creatinine and histidine in Benedict's qualitative test for reducing sugar in urine. J. Clin. Pathol. 25, 892–895. https://doi.org/10.1136/jcp.25.10.892.

Syal, K., Banerjee, D., Srinivasan, A., 2013. Creatinine estimation and interference. Indian J. Clin. Biochem. 28, 210–211. https://doi.org/10.1007/s12291-013-0299-y.

Thyagaraju, K., Sujatha, G., Venkataswamy, M., Bukke, S., Katepogu, K., 2016. Serum glutamate pyruvate transaminase is best marker enzyme for thyroid function—a report based on Nellore females, Andhra Pradesh, India. Int. J. Adv. Res. 4, 777–784.

Toora, B.D., Rajagopal, G., 2002. Measurement of creatinine by Jaffe's reaction—determination of concentration of sodium hydroxide required for maximum color development in standard, urine and protein free filtrate of serum. Indian J. Exp. Biol. 40, 352–354.

Van Noorden, R., Maher, B., Nuzzo, R., 2014. The top 100 papers. Nature 514, 550–553. https://doi.org/10.1038/514550a.

Watson, D., 1960. A simple method for the determination of serum cholesterol. Clin. Chim. Acta 5, 637–643. https://doi.org/10.1016/0009-8981(60)90004-8.

Watts, R.W.E., 1974. Determination of uric acid in blood and in urine. Ann. Clin. Biochem. 11, 103–111. https://doi.org/10.1177/000456327401100139.

Wiechelman, K.J., Braun, R.D., Fitzpatrick, J.D., 1988. Investigation of the bicinchoninic acid protein assay: identification of the groups responsible for color formation. Anal. Biochem. 175, 231–237. https://doi.org/10.1016/0003-2697(88)90383-1.

Woods, J., Mellon, M., 1941. Thiocyanate method for iron: a spectrophotometric study. Ind. Eng. Chem. Anal. Ed. 13, 551–554. https://doi.org/10.1021/i560096a013.

Xiong, Q., Wilson, W.K., Pang, J., 2007. The Liebermann-Burchard reaction: sulfonation, desaturation, and rearrangement of cholesterol in acid. Lipids 42, 87–96. https://doi.org/10.1007/s11745-006-3013-5.

Yanagisawa, F., 1955. New colorimetric determination of calcium and magnesium. J. Biochem. 42, 3–11. https://doi.org/10.1093/oxfordjournals.jbchem.a126504.

Zhao, Y., Yang, X., Lu, W., Liao, H., Liao, F., 2009. Uricase based methods for determination of uric acid in serum. Microchim. Acta 164, 1–6. https://doi.org/10.1007/s00604-008-0044-z.

Book

AOAC, 1984. In: Harwitz, W. (Ed.), Official Methods of Analysis, thirteenth ed. Association of Official Analytical Chemists, Washington, DC.

Arnon, R., 1970. In: Perlmann, G., Lorand, L. (Eds.), Methods in Enzymol. vol. 19. Academic Press, New York, p. 228.

Ausubel, F.M., Brent, R., Kingston, R.E., Moore, D.D., Seidman, J.G., Smith, J.A., Struhl, K., 1987. Current Protocols in Biochemistry. Wiley, New York.

Baker, F., 1988. Determination of Serum Tocopherol by Colorimetric Method. Varley' s Practical Clinical Biochemistry, sixth ed. Heinemann Professional Publishing, Portsmouth, NH, USA, p. 902.

Bowen, J.B., Graham, S.H., Williams, A.I.S., 1957. A Students' Handbook of Organic Qualitative Analysis. University of London Press, London, UK.

Chawla, R., 2014. Practical Clinical Biochemistry: Methods and Interpretations. Jaypee Brothers Medical Publishers (P) Ltd, New Delhi, India. ISBN: 978-81-8061-108-7.

Conn, E.E., Stumpf, P.K., 1987. Outlines of Biochemistry, fourth ed. Wiley Eastern, New Delhi, pp. 3–24.

Gupta, A., 2019. "Carbohydrates". Comprehensive Biochemistry for Dentistry. Springer, Singapore, ISBN: 978-981-13-1035-5, pp. 108–110.

Jain, A., Jain, R., Jain, S., 2020. Paper chromatography of amino acid. In: Basic Techniques in Biochemistry, Microbiology and Molecular Biology. Springer Protocols Handbooks. Humana, New York, NY, ISBN: 978-1-4939-9861-6, https://doi.org/10.1007/978-1-4939-9861-6_59.

Kumar, V., Gill, K.D., 2018. Basic Concepts in Clinical Biochemistry: A Practical Guide. Springer, Singapore, ISBN: 978-981-10-8186-6, https://doi.org/10.1007/978-981-10-8186-6_26.

Longo, D.L., et al., 2016. Harrison's Principles of Internal Medicine, nineteenth ed. McGraw-Hill Education, New York, NY, ISBN: 978-0-07-180215-4.

Mahajan, R., 2013. Manual of Practical Biochemistry for Dental Students. JBC Press, ISBN: 9789382174943.

Maniatis, T., Fritsch, E.F., Sambrook, J., 1982. Molecular Cloning—A Laboratory Manual. Cold Spring Harbour Laboratory, New York. ISBN:0-87969-136-0.

Ninfa, A.J., 2010. Fundamental Laboratory Approaches for Biochemistry and Biotechnology. John Wiley & Sons, USA, ISBN: 978-0-470-08766-4, p. 112.

Plummer, D.T., 1971. An Introduction of Practical Biochemistry. Tata McGraw-Hill, New Delhi, ISBN: 978-0070994874, p. 228.

Racek, J., Rajdi, D., 2016. Clinical Biochemistry. Karolinum Press. Charles University, Prague.

Ramakrishnan, S., 2004. Textbook of Medical Biochemistry. Orient Blackswan, ISBN: 9788125020714.

Robert, J.D., Caserio, M.C., 1977. Basic Principles of Organic Chemistry, second ed. W.A. Benjamin, Menlo Park, CA, ISBN: 0-8053-8329-8.

Sambrook, J., Fritsch, E.F., Maniatis, T., 1989. Molecular Cloning: A Laboratory Manual. vol. 3 Cold Spring Harbour Laboratory Press, New York.

Segel, I.H., 1976. Biochemical Calculations, second ed. Wiley, New York.

Slater, R.J., 1986. Experiments in Molecular Biology. Humana Press, Clifton, NJ, ISBN: 978-1-60327-405-0, p. 269.

Smith, B.J., 1984. SDS polyacrylamide gel electrophoresis of proteins. Methods Mol. Biol. 1, 41–56. https://doi.org/10.1385/0-89603-062-8:41. ISBN:0-89603-062-8. 20512673.

Vasudevan, D.M., Das, S.K., 2013. Practical Textbook of Biochemistry for Medical Students. Jaypee Brothers Medical Publishers, ISBN: 9789389034981.

Walsh, E.O.'.F., 1961. An Introduction to Biochemistry. The English Universities Press, London, pp. 406–407. OCLC 421450365.

West, E.S., Todd, W.R., Mason, H.S., Van Bruggen, J.T., 1966. Textbook of Biochemistry, fourth ed. Macmillan, New York, USA.

Wilson, K., Walker, J., 2013. Principles and Techniques of Biochemistry and Molecular Biology, seventh ed. Cambridge University Press, ISBN: 9780511841477.

Websites and Other Sources

Anon. http://dept.harpercollege.edu/chemistry/chm/100/dgodambe/thedisk/carbo/bial/bials.htm.

Anon. http://egyankosh.ac.in/.

Anon. http://generalchemistrylab.blogspot.com/2011/12/mucic-acid-test-for-galactose.html.

Anon. https://alevelbiology.co.uk/notes/tests-for-carbohydrates/.

Anon. https://microbiologyinfo.com/.

Anon. https://shodhganga.inflibnet.ac.in/.

Anon. https://vlab.amrita.edu.

Anon. https://websys.knust.edu.gh/oer/pages/index.php?siteid=knustoer&page=find_materials&cou=38.

Anon. https://www.academia.edu.

Anon. https://www.chemistrylearner.com/benedicts-test.html.

Index

Note: Page numbers followed by *f* indicate figures and *t* indicate tables.

Printed in the United States
By Bookmasters